SPACE SCIENCE, EXPLORATION AND POLICIES SERIES

SUSTAINING THE GLOBAL POSITIONING SYSTEM

SPACE SCIENCE, EXPLORATION AND POLICIES SERIES

Progress in Dark Matter Research
J. Val Blain
(Editor)
2005. ISBN 1-59454-248-1

Space Science: New Research
Nick S. Maravell
(Editor)
2006. ISBN 1-60021-005-8

Space Policy and Exploration
William N. Callmers (Editor)
2008. ISBN 978-1-60456-448-8

Space Commercialization and the Development of Space Law from a Chinese Legal Perspective
Yun Zhao
2009. ISBN 978-1-60692-244-6

Space Commercialization and the Development of Space Lawfrom a Chinese Legal Perspective
Yun Zhao
2009. ISBN 978-1-60876-874-5
(Online Book)

Space Exploration Research
John H. Denis and Paul D. Aldridge
(Editors)
2009. ISBN: 978-1-60692-264-4

Space Exploration Research
John H. Denis and Paul D. Aldridge
(Editors)
2009. ISBN: 978-1-61668-212-5
(Online Book)

Next Generation of Human Space Flight Systems
Alfred T. Chesley (Editor)
2009. ISBN 978-1-60692-726-7

Nutritional Biochemistry of Space Flight
Scott M. Smith, Sara R. Zwart, Vickie Kloeris and Martina Heer
2009. ISBN 978-1-60741-641-8

Environmental Satellites: Weather and Environmental Information Systems
Vincent L. Webber (Editor)
2009. ISBN 978-1-60692-984-1

Smaller Satellites Operations Near Geostationary Orbit
Matthew T. Erdner
2009. ISBN 978-1-60741-181-9

Black Holes and Galaxy Formation
Adonis D. Wachter and Raphael J. Propst
(Editors)
2010. ISBN: 978-1-60741-703-3

Dark Energy: Theories, Developments, and Implications
Karl Lefebvre and Raoul Garcia (Editors)
2010. ISBN: 978-1-61668-271-2

Dark Energy: Theories, Developments, and Implications
Karl Lefebvre and Raoul Garcia (Editors)
2010. ISBN: 978-1-61668-889-9 (Online Book)

Future U.S. Space Launch Capabilities
Franz Lojdahl (Editor)
2010. ISBN: 978-1-60741-384-4

The International Space Station
David E. Jamison (Editor)
2010. ISBN: 978-1-60692-322-1

Global Positioning Systems
Viggo Asphaug and Elias Sørensen (Editors)
2010. ISBN: 978-1-60741-012-6

Astrobiology: Physical Origin, Biological Evolution and Spatial Distribution
Simon Hegedűs and Jakob Csonka (Editors)
2010. ISBN: 978-1-60741-290-8

Space Tourism Issues
Elias Wikborg (Editor)
2010. ISBN: 978-1-60741-353-0

Exploring the Final Frontier : Issues, Plans and Funding for NASA
Dillon S. Maguire (Editor)
2010. ISBN: 978-1-60876-080-0

Primordial Space
Bernd Schmeikal
2010. ISBN: 978-1-60876-781-6

Sustaining the Global Positioning System
Earl M. Peabody (Editor)
2010. ISBN: 978-1-60741-006-5

Analytical Methods for the Formation of Dark Matter Haloes in the Universe
Nicos Hiotelis
2010. ISBN: 978-1-60876-473-0

Space Material Sciences
A.I. Feonychev
2010. ISBN: 978-1-61668-236-1

Space Material Sciences
A.I. Feonychev
2010. ISBN: 978-1-61668-477-8 (Online Book)

Commercial Space Transportation
Jocelyn S. Gunther (Editor)
2010. ISBN: 978-1-61668-707-6

Commercial Space Transportation
Jocelyn S. Gunther (Editor)
2010. ISBN: 978-1-61668-795-3 (Online Book)

Future of U.S. Human Spaceflight: Background and Issues

Derek A. Warren and Bridget D. Conway (Editors)

2010. ISBN: 978-1-61668-774-8

Future of U.S. Human Spaceflight: Background and Issues

Derek A. Warren and Bridget D. Conway (Editors)

2010. ISBN: 978-1-61668-876-9 (Online Book)

How Did We All Begin: Where is God in All That?

Calvin S. Kalman

2010. ISBN:978-1-61668-364-1

SPACE SCIENCE, EXPLORATION AND POLICIES SERIES

SUSTAINING THE GLOBAL POSITIONING SYSTEM

EARL M. PEABODY
EDITOR

Nova Science Publishers, Inc.
New York

For permission to use material from this book please contact us:
Telephone 631-231-7269; Fax 631-231-8175
Web Site: http://www.novapublishers.com

LIBRARY OF CONGRESS CATALOGING-IN-PUBLICATION DATA

Sustaining the global positioning system / editor, Earl M. Peabody.
p. cm.
Includes index.
ISBN 978-1-60741-006-5 (softcover)
1. Global Positioning System. I. Peabody, Earl M.
G109.5.S87 2009
623.89'3--dc22

2009038417

Published by Nova Science Publishers, Inc. ✢ New York

CONTENTS

PREFACE

This book explores the significant challenges in sustaining and upgrading The Global Positioning System (GPS). The GPS provides positioning, navigation, and timing data to users worldwide, and has become essential to U.S. national security and a key tool in an expanding array of public service and commercial applications at home and abroad. GPS is integrated into nearly every facet of U.S. military operations, and the number of civil users is increasing. Other countries are now developing their own independent global navigation satellite systems that could offer capabilities that are comparable, if not superior to GPS. The U.S. government, which plans to invest more than $5.8 billion from 2009 through 2013 in the GPS space and ground control segments currently under development, provides GPS service free of charge. The Department of Defense (DoD) develops and operates GPS, and an interdepartmental committee manages the U.S. space-based positioning, navigation and timing infrastructure, which includes GPS. This book looks at the global economic and national security importance of GPS, the ongoing GPS modernization effort and the international efforts to develop new systems.

Chapter 1 - The Global Positioning System (GPS), which provides positioning, navigation, and timing data to users worldwide, has become essential to U.S. national security and a key tool in an expanding array of public service and commercial applications at home and abroad. The United States provides GPS data free of charge. The Air Force, which is responsible for GPS acquisition, is in the process of modernizing GPS.

In light of the importance of GPS, the modernization effort, and international efforts to develop new systems, GAO was asked to undertake a broad review of GPS. Specifically, GAO assessed progress in (1) acquiring GPS satellites, (2) acquiring the ground control and user equipment necessary to leverage GPS satellite capabilities, and evaluated (3) coordination among federal agencies and other organizations to ensure GPS missions can be accomplished. To carry out this assessment, GAO's efforts included reviewing and analyzing program documentation, conducting its own analysis of Air Force satellite data, and interviewing key military and civilian officials.

Chapter 2 - This chapter is edited and excerpted testimony Mr. Chet Huber, before the Subcommittee on National Security and Foreign Affairs, Committee on Oversight and Government Reform, House of Representatives, on May 7, 2009.

Chapter 3 - This chapter is edited and excerpted testimony Lieutenant General Larry James, before the Subcommittee on National Security and Foreign Affairs, Committee on Oversight and Government Reform, House of Representatives, on May 7, 2009.

Chapter 4 - This chapter is edited and excerpted testimony Major General Neil McCasland, before the Subcommittee on National Security and Foreign Affairs, Committee on Oversight and Government Reform, House of Representatives, on May 7, 2009.

Chapter 5 - This chapter is edited and excerpted testimony Mr. F. Michael Swiek, before the Subcommittee on National Security and Foreign Affairs, Committee on Oversight and Government Reform, House of Representatives, on May 7, 2009.

Chapter 6 - This chapter is edited and excerpted testimony Karen Van Dyke, before the Subcommittee on National Security and Foreign Affairs, Committee on Oversight and Government Reform, House of Representatives, on May 7, 2009.

In: Sustaining the Global Positioning System ISBN: 978-1-60741-006-5
Editors: Earl M. Peabody

Chapter 1

GLOBAL POSITIONING SYSTEM: SIGNIFICANT CHALLENGES IN SUSTAINING AND UPGRADING WIDELY USED CAPABILITIES

Government Accountability Office

WHY GAO DID THIS STUDY

The Global Positioning System (GPS), which provides positioning, navigation, and timing data to users worldwide, has become essential to U.S. national security and a key tool in an expanding array of public service and commercial applications at home and abroad. The United States provides GPS data free of charge. The Air Force, which is responsible for GPS acquisition, is in the process of modernizing GPS.

In light of the importance of GPS, the modernization effort, and international efforts to develop new systems, GAO was asked to undertake a broad review of GPS. Specifically, GAO assessed progress in (1) acquiring GPS satellites, (2) acquiring the ground control and user equipment necessary to leverage GPS satellite capabilities, and evaluated (3) coordination among federal agencies and other organizations to ensure GPS missions can be accomplished. To carry out this assessment, GAO's efforts included reviewing and analyzing program documentation, conducting its own analysis of Air Force satellite data, and interviewing key military and civilian officials.

What GAO Recommends

GAO's recommendations include that the Secretary of Defense appoint a single authority to oversee development of GPS space, ground control, and user equipment assets, to ensure they are synchronized, well executed, and potential disruptions are minimized. DOD concurred with our recommendations.

What GAO Found

It is uncertain whether the Air Force will be able to acquire new satellites in time to maintain current GPS service without interruption. If not, some military operations and some civilian users could be adversely affected.

- In recent years, the Air Force has struggled to successfully build GPS satellites within cost and schedule goals; it encountered significant technical problems that still threaten its delivery schedule; and it struggled with a different contractor. As a result, the current IIF satellite program has overrun its original cost estimate by about $870 million and the launch of its first satellite has been delayed to November 2009—almost 3 years late.
- Further, while the Air Force is structuring the new GPS IIIA program to prevent mistakes made on the IIF program, the Air Force is aiming to deploy the next generation of GPS satellites 3 years faster than the IIF satellites. GAO's analysis found that this schedule is optimistic, given the program's late start, past trends in space acquisitions, and challenges facing the new contractor. Of particular concern is leadership for GPS acquisition, as GAO and other studies have found the lack of a single point of authority for space programs and frequent turnover in program managers have hampered requirements setting, funding stability, and resource allocation.
- If the Air Force does not meet its schedule goals for development of GPS IIIA satellites, there will be an increased likelihood that in 2010, as old satellites begin to fail, the overall GPS constellation will fall below the number of satellites required to provide the level of GPS service that the U.S. government commits to. Such a gap in capability could have wide-ranging impacts on all GPS users, though there are measures the Air Force and others can take to plan for and minimize these impacts.

In addition to risks facing the acquisition of new GPS satellites, the Air Force has not been fully successful in synchronizing the acquisition and development of the next generation of GPS satellites with the ground control and user equipment, thereby delaying the ability of military users to fully utilize new GPS satellite capabilities. Diffuse leadership has been a contributing factor, given that there is no single authority responsible for synchronizing all procurements and fielding related to GPS, and funding has been diverted from ground programs to pay for problems in the space segment.

DOD and others involved in ensuring GPS can serve communities beyond the military have taken prudent steps to manage requirements and coordinate among the many organizations involved with GPS. However, GAO identified challenges to ensuring civilian requirements and ensuring GPS compatibility with other new, potentially competing global space-based positioning, navigation, and timing systems.

ABBREVIATIONS

AEP	Architecture Evolution Plan
DASS	Distress Alerting Satellite System
DOD	Department of Defense
GNSS	Global Navigation Satellite Systems
GPS	Global Positioning System
IFOR	Interagency Forum for Operational Requirements
JCIDS	Joint Capabilities Integration and Development System
L2C	second civil signal
L5	third civil signal
M-code	Military Code
NASA	National Aeronautics and Space Administration
OCS	Operational Control Segment
OCX	Next Generation Control Segment
OSD	Office of the Secretary of Defense
PDOP	position dilution of precision
PNT	Positioning, Navigation, and Timing
SLR	Satellite Laser Ranging
TSPR	Total System Performance Responsibility

April 30, 2009

The Honorable John Tierney
Chairman
The Honorable Jeff Flake Ranking Member
Subcommittee on National Security and Foreign Affairs
Committee on Oversight and Government Reform
House of Representatives

The Global Positioning System (GPS)—a space-based satellite system that provides positioning, navigation, and timing data to users worldwide—has become essential to U.S. national security and a key component in economic growth, transportation safety, homeland security, and critical national infrastructure in the United States and abroad. GPS is integrated into nearly every facet of U.S. military operations, and the number of civil users is increasing. Other countries are now developing their own independent global navigation satellite systems that could offer capabilities that are comparable, if not superior to GPS.

The U.S. government, which plans to invest more than $5.8 billion from 2009 through 2013 in the GPS space and ground control segments currently under development, provides GPS service free of charge. The Department of Defense (DOD) develops and operates GPS, and an interdepartmental committee—co-chaired by DOD and the Department of Transportation—manages the U.S. space-based positioning, navigation, and timing infrastructure, which includes GPS. DOD also provides most of the funding for GPS.

The Air Force, which is responsible for GPS acquisition, is in the process of modernizing GPS to enhance its performance, accuracy, and integrity. The modernization effort includes GPS IIF and IIIA, two satellite acquisition programs currently underway that are to provide new space- based capabilities and replenish the satellite constellation; the ground control segment hardware and software; and user equipment for processing modernized GPS capabilities.

In light of the global economic and national security importance of GPS, the ongoing GPS modernization effort, and the international efforts to develop new systems, you asked us to undertake a broad review of the program and efforts to replenish and upgrade capability. Specifically, we assessed progress in (1) acquiring GPS satellites, (2) acquiring the ground control and user equipment necessary to leverage GPS satellite capabilities, and (3) coordinating among federal agencies and other organizations to ensure broader GPS missions can be accomplished.

To assess the acquisition of satellite, ground control, and user equipment, we interviewed Office of the Secretary of Defense (OSD) and DOD officials from offices that manage and oversee the GPS program. We also reviewed and analyzed program plans and documentation related to cost, schedule, requirements, program direction, and satellite constellation sustainment, and compared programmatic data to GAO's criteria compiled over the last 12 years for best practices in system development.[1] We also conducted our own analysis, based on data provided by the Air Force, to assess the implications of potential schedule delays we identified in our assessment of the satellite acquisition. To assess coordination among federal agencies and the broader GPS community, we interviewed OSD and DOD officials from offices that manage and oversee the GPS program, officials from the military services, officials from the Department of Transportation and other civil departments and agencies, and officials at the U.S. Department of State and at various European space organizations. We also analyzed how civil departments and agencies coordinate with DOD on GPS civil requirements, and how the U.S. government coordinates with foreign countries. Additional information on our scope and methodology is in appendix I. We conducted this performance audit from October 2007 to April 2009 in accordance with generally accepted government auditing standards. Those standards require that we plan and perform the audit to obtain sufficient, appropriate evidence to provide a reasonable basis for our findings and conclusions based on our audit objectives. We believe that the evidence obtained provides a reasonable basis for our findings and conclusions based on our audit objectives.

RESULTS IN BRIEF

It is uncertain whether the Air Force will be able to acquire new satellites in time to maintain current GPS service without interruption. If not, some military operations and some civilian users could be adversely affected.

- Under the IIF program, the Air Force had difficulty in successfully building GPS satellites within cost and schedule goals; it encountered significant technical problems which still threaten its delivery schedule; and it faced challenges with a different contractor for the IIF program. These problems were compounded by an acquisition strategy that relaxed oversight and quality inspections as well as multiple contractor mergers and moves, and the addition of new requirements late in the development cycle. As a result, the IIF program has overrun its original cost estimate of $729 million by about $870 million and the launch of the first IIF satellite has been delayed to November 2009—almost 3 years late.

- Further, while the Air Force is structuring the new GPS IIIA program to prevent mistakes made on the IIF program, the Air Force is aiming to deploy the GPS IIIA satellites 3 years faster than the IIF satellites. We believe the IIIA schedule is optimistic given the program's late start, past trends in space acquisitions, and challenges facing the new contractor. Of particular concern is leadership for GPS acquisition, as GAO and other studies have found the lack of a single point of authority for space programs and frequent turnover in program managers have hampered requirements setting, funding stability, and resource allocation.
- If the Air Force does not meet its schedule goals for development of GPS IIIA satellites, there will be an increased likelihood that in 2010, as old satellites begin to fail, the overall GPS constellation will fall below the number of satellites required to provide the level of GPS service that the U.S. government is committed to providing. Such a gap in capability could have wide-ranging impacts on all GPS users, though there are measures the Air Force and others can take to plan for and minimize these impacts.

Moreover, the Air Force has not been fully successful in synchronizing the acquisition and development of the next generation of GPS satellites with the ground control and user equipment, thereby delaying the ability of military users to utilize new GPS satellite capabilities. For example, a modernized military signal will be available for operations on GPS satellites over a decade before user equipment will be fielded that can take strategic advantage of it. The signal is designed to improve resistance to jamming of GPS. Also, because leadership for acquisitions across the space community is fragmented, there is no single authority responsible for synchronizing all procurements and fielding related to GPS.

Lastly, DOD and others involved in ensuring GPS can serve communities beyond the military have taken prudent steps to manage requirements and coordinate among the many organizations involved with GPS. However, we identified challenges in the areas of ensuring civilian requirements can be met and ensuring GPS compatibility with other new, potentially competing global space-based positioning, navigation, and timing systems.

Because of (1) the criticality of the GPS system to the military, various economic sectors, and the international community and (2) schedule risks in the current program, we are recommending that the Secretary of Defense appoint a single authority to oversee the development of the GPS system, including DOD space, ground control, and user equipment assets, to ensure that the program is well executed and resourced and that potential disruptions are minimized. The appointee should have authority to ensure DOD space, ground control, and user equipment are synchronized to the maximum extent practicable; and coordinate with the existing positioning, navigation, and timing infrastructure to assess and minimize potential service disruptions in the event that the satellite constellation was to decrease in size for an extended period of time. After a review of a draft of this report, DOD concurred with our recommendations and provided some additional comments. The full text of DOD's comments may be found in appendix IV.

BACKGROUND

GPS is a global positioning, navigation, and timing network consisting of space, ground control, and user equipment segments that support the broadcasts of military and civil GPS signals. These signals each include positioning and timing information, which enables users with GPS receivers to determine their position, velocity, and time, 24 hours a day, in all weather, worldwide. GPS is used by all branches of the military to guide troops' movements, integrated logistics support and battlespace situational awareness, and communications network synchronization. In addition, bombs and missiles are guided to their targets by GPS signals and GPS is used to locate military personnel in distress. Early in the development of GPS, the scope was expanded to include complementary civil capabilities.

Over time, GPS has become a ubiquitous infrastructure underpinning major sections of the economy, including telecommunications, electrical power distribution, banking and finance, transportation, environmental and natural resources management, agriculture, and emergency services in addition to the array of military operations it services. For instance, civil agencies, commercial firms, and individuals use GPS to accurately navigate from one point to another. Commercial firms use GPS to route their vehicles, as do maritime industries and mass transit systems. In addition to navigation, civil departments and agencies and commercial firms use GPS and GPS augmentations[2] to provide high-accuracy, three-dimensional positioning information in real time for use in surveying and mapping. The aviation community worldwide uses GPS and GPS augmentations to increase the safety and efficiency of flight. GPS is also used in the agricultural community for precision farming, including farm planning, field mapping, soil sampling, tractor guidance, and crop scouting. GPS helps companies and governments place satellites in precise orbits, and at correct altitudes, and helps monitor satellite constellation orbits. The precise time that GPS broadcasts is crucial to economic activities worldwide, including communication systems, electrical power grids, and financial networks.

GPS System Description

GPS operations consist of three segments—the space segment, the ground control segment, and the user equipment segment. All segments are needed to take full advantage of GPS capabilities.

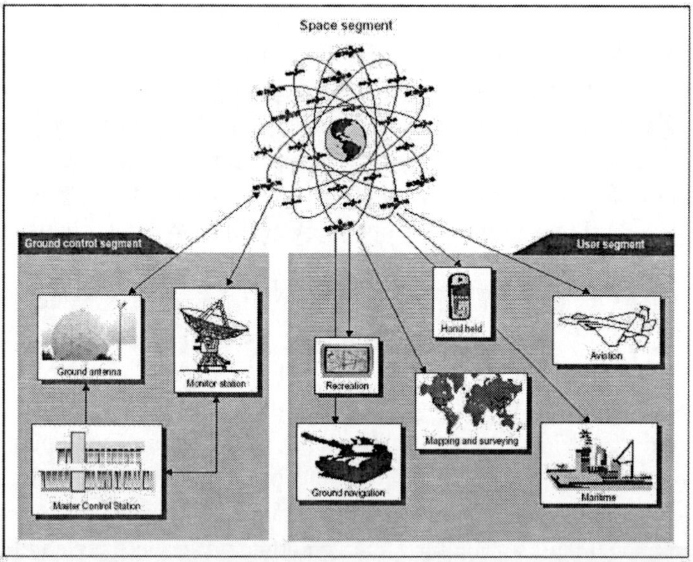

Source: Copyright © Corel Corp. All rights reserved (map); Art Explosion; GAO.

Figure 1. GPS Operational System

The GPS space segment consists of a constellation of satellites that move in six orbital planes approximately 20,200 kilometers above the earth. GPS satellites broadcast encrypted military signals and civil signals. In recent years, because numerous satellites have exceeded their design life, the constellation has grown to 31 active satellites of various generations.

Figure 2 describes the National Space-Based PNT organization structure.

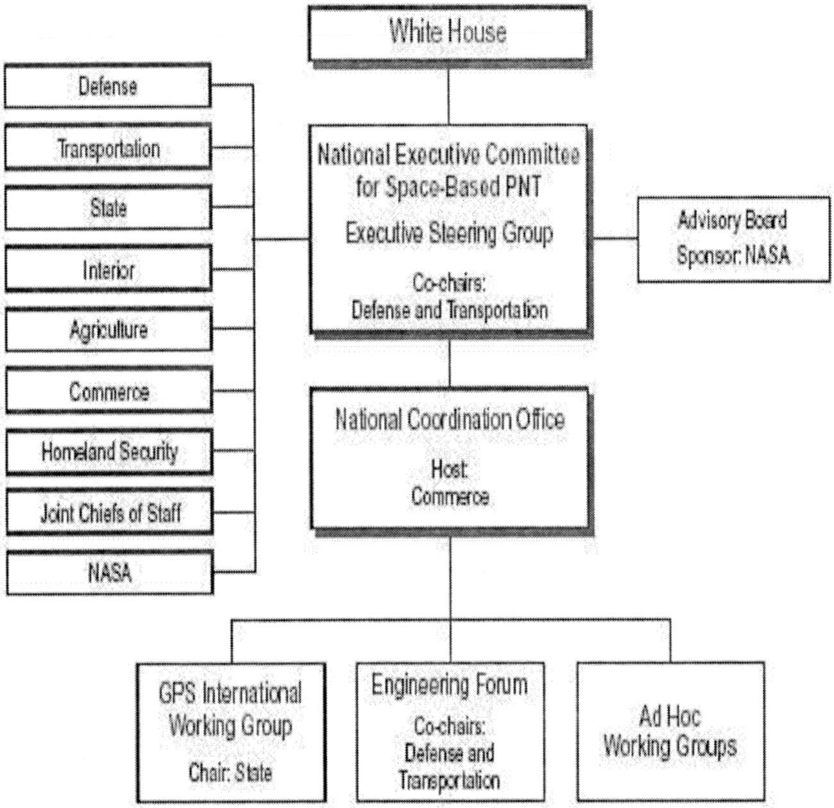

Source: GAO presentation of National Executive Committee for Space-Based Positioning, Navigation and Timing Data.

Figure 2. National Space-Based PNT Organization Structure

However, DOD predicts that over the next several years many of the older satellites in the constellation will reach the end of their operational life faster than they will be replenished, thus decreasing the size of the constellation from its current level and potentially reducing the accuracy of the GPS service

The GPS ground control segment is comprised of a Master Control Station at Schriever Air Force Base, Colorado; an Alternate Master Control Station at

Vandenberg Air Force Base, California; 6 Air Force and 11 National Geospatial-Intelligence Agency monitoring stations; and four ground antennas with uplink capabilities. Information from the monitoring stations is processed at the Master Control Station to determine satellite clock and orbit status. The Master Control Station operates the satellites and regularly updates the navigation messages on the satellites. Information from the Master Control Station is transmitted to the satellites via the ground antennas.

The GPS user equipment segment includes military and commercial GPS receivers. These receivers determine a user's position by calculating the distance from four or more satellites using the navigation message on the satellites to triangulate its location. Military GPS receivers are designed to utilize the encrypted military GPS signals that are only available to authorized users, including military and allied forces and some authorized civil agencies. Commercial receivers use the civil GPS signal, which is publicly available worldwide.

Full use of the military and civil GPS signals requires a ground control system that can manage these signals. Newer software will upgrade the ground control to a service oriented—or "plug and play"—architecture that can connect to broader networks. In order to utilize the modernized military signal from the ground, military users require new user equipment with this capability, which will be provided by the military GPS user equipment program.

Broader Management Structure

The 2004 U.S. Space-Based Positioning, Navigation and Timing (PNT) policy established a management structure to bring civil and military departments and agencies together to form an interagency, multiuse approach to program planning, resource allocation, system development, and operations. The policy also encourages cooperation with foreign governments to promote the use of civil aspects of GPS and its augmentation services and standards with foreign governments and other international organizations. As part of the management structure, an executive committee advises and coordinates among U.S. government departments and agencies on maintaining and improving U.S. space-based PNT infrastructures, including GPS and related systems. The executive committee is co-chaired by the Deputy Secretaries of the Department of Defense and the Department of Transportation, and includes members at the equivalent level from the Departments of State, Commerce, Homeland Security, Interior, Agriculture, the Joint Chiefs of Staff, and the National Aeronautics and Space

Administration (NASA). The departments and agencies have various assigned roles and responsibilities. For example, DOD is responsible for the overall development, acquisition, operation, security, and continued modernization of GPS. It has delegated acquisition responsibility to the Air Force, though other DOD components and military services are responsible for oversight, some aspects of user equipment development, and for funding some parts of the program. The Department of Transportation has the lead responsibility for the coordination of civil requirements from all civil department and agencies. The Department of State leads negotiations with foreign governments and international organizations on GPS positioning, navigation, and timing matters or regarding the planning, operations, management, and/or use of GPS. (See app. III).

Broader Management Structure

The 2004 U.S. Space-Based Positioning, Navigation and Timing (PNT) policy established a management structure to bring civil and military departments and agencies together to form an interagency, multiuse approach to program planning, resource allocation, system development, and operations. The policy also encourages cooperation with foreign governments to promote the use of civil aspects of GPS and its augmentation services and standards with foreign governments and other international organizations. As part of the management structure, an executive committee advises and coordinates among U.S. government departments and agencies on maintaining and improving U.S. space-based PNT infrastructures, including GPS and related systems. The executive committee is co-chaired by the Deputy Secretaries of the Department of Defense and the Department of Transportation, and includes members at the equivalent level from the Departments of State, Commerce, Homeland Security, Interior, Agriculture, the Joint Chiefs of Staff, and the National Aeronautics and Space Administration (NASA). The departments and agencies have various assigned roles and responsibilities. For example, DOD is responsible for the overall development, acquisition, operation, security, and continued modernization of GPS. It has delegated acquisition responsibility to the Air Force, though other DOD components and military services are responsible for oversight, some aspects of user equipment development, and for funding some parts of the program. The Department of Transportation has the lead responsibility for the coordination of civil requirements from all civil department and agencies. The Department of State leads negotiations with foreign governments and international organizations on GPS positioning, navigation, and timing matters or regarding the planning, operations, management, and/or use of GPS. (See app. III). Table 1 shows the modernization efforts for the space and ground control segment.

Table 1. GPS Satellite and Ground Control Segment Modernization

Satellite evolution and capabilities			
Legacy (1989 - 2002)		**Current (2005 - 2012)**	**Future (2014 - 2023)**
GPS IIA/IIR	**GPS IIR-M**	**GPS IIF**	**GPS III**
• Broadcasts signals for • military and civil users.	Includes IIA and IIR capabilities, plus: • 2nd civil signal • 2nd military signal • Ability to increase signal power to improve resistance to jamming	Includes IIR-M capabilities, plus: • 3rd civil signal for Transportation safety requirements	Includes IIF capabilities, plus: • IIIA: stronger military signal to improve jamming resistance and 4th civil signal that is interoperable with foreign signals • IIIB: near real-time command and control via cross links • IIIC: improved antijam performance for military users
Ground control segment and capabilities			
Legacy (Various versions from 1979-2007)	**Current (Came online in 2007)**		**Future (Planned to come online with initial capabilities in 2011)**
Legacy Operational Control System (OCS)	**Architecture Evolution Plan (AEP)**		**Next Generation Operational Control Segment (OCX)**
• Centralized computer mainframe • 1970s technology	• Distributed architecture • Enables upgrades to the system • Next upgrade will control GPS IIF		• Necessary for full operation of GPS IIR-M, IIF, and III satellites • Service-oriented architecture • Connects to the broader network

Source: GAO analysis based on DOD program information and discussions with DOD officials.

GPS Modernization

In 2000, DOD began an effort to modernize the space, ground control, and user equipment segments of GPS to enhance the system's performance, accuracy, and integrity.

AIR FORCE FACES SIGNIFICANT CHALLENGES IN ACQUIRING GPS SATELLITES

The Air Force's GPS IIF acquisition initially was not well executed, and currently poses technical problems. The Air Force is implementing lessons learned from the GPS IIF effort as it starts the GPS IIIA program. However, based on our analysis, the GPS IIIA program faces a compressed schedule along with new challenges to deliver the satellites on time. A slip in the launch of the GPS IIIA satellites could increase the likelihood that the GPS constellation will fall below the number of satellites required to provide the level of GPS service the U.S. government has committed to provide. This would not only have implications for military users but also for the larger community of GPS users, who may be less aware and equipped to deal with gaps in coverage. However, the Air Force is evaluating different approaches that could potentially reduce the risk of degrading the GPS service.

The IIF Program Was Not Well Executed, and Still Poses Technical Problems

The GPS IIF contract was awarded during an era of acquisition reform that centered on an approach called Total System Performance Responsibility (TSPR).[3] TSPR gave a contractor total responsibility for the integration of an entire weapon system and for meeting DOD's requirements. This approach was intended to facilitate acquisition reform and enable DOD to streamline a cumbersome acquisition process and leverage innovation and management expertise from the private sector. However, DOD later found that TSPR magnified problems on a number of satellite acquisition programs because it was implemented in a manner that enabled requirements creep and poor contractor performance. For GPS IIF, the TSPR approach resulted in relaxed specifications and inspections of the contractor, loss of quality in the manufacturing process, and poor-quality parts that caused test failures, unexpected redesigns, and the late delivery of parts. The contractor did not provide data on design drawings and statistical process control techniques were not used to monitor production.

According to GPS program officials, the GPS IIF program was also negatively impacted by multiple contractor mergers, acquisitions, and moves. In 1996, shortly after Rockwell won the IIF contract, the company's aerospace and defense units, including the Seal Beach, California, facility where the IIF satellites were to be manufactured, were acquired by Boeing. In December 1997, Boeing merged with McDonnell Douglas and took over its Delta launch vehicle unit in Huntington Beach, California, and subsequently GPS work was moved to that facility. In October 2000, Boeing acquired Hughes Electronics Corporation's space and communications business and related operations. Boeing took over the Hughes facility in El Segundo, California, and once again, GPS work was moved to another facility. As these events occurred, the prime contractor consolidated development facilities to remain competitive. In addition, the prime contractor lost valuable workers and knowledge, causing inefficiencies in the program.

Shortly after the IIF contract was awarded in 1996, the Air Force also added requirements. For example, the government decided to accelerate the fielding of new civil and military GPS signals. Flexible power capabilities were added to IIF several years later. These new requirements drove design changes and resulted in technical issues and cost overruns that impacted the schedule.

According to a GPS IIF program official, the combination of significant requirements additions, loss of engineering expertise, parts obsolescence, and fundamental design changes together caused the contractor to "lose the recipe" for the IIF space vehicle. In essence, by the completion of the design phase, the IIF space vehicle was to be built in a third location, by different people, in a way that was not initially anticipated. In addition, the program suffered from a lack of management continuity. Since the program's inception, the IIF program has had seven different program managers, the first five of whom only served 1 year each.

According to a former deputy program director of the GPS program office, past GPS programs seemed to operate well for a number of reasons. The programs (1) never added major modifications to ongoing programs and (2) had no qualms in terminating contractors if work did not meet standards, business practices, or major milestones. Furthermore, the GPS program performed more on-site contract management to increase communications. This approach eliminated surprises like cost and schedule overruns and held the contractor to a high level of performance. Lastly, the former director noted that it was important to balance the responsibility assigned to the program managers with the authority they needed to properly implement the program. Prior GAO reviews have identified all of these practices as essential to program execution.[4]

Air Force Improves Oversight of IIF, but Technical Issues Lead to More Delays

The Air Force has since taken action to improve the IIF program. In 2006, the program office[5] increased its personnel at the contractor's facility to observe operations and to verify that corrective measures were being taken to address deficiencies in the contractor's cost and schedule reporting system (also known as earned value management[6]). The Air Force increased the number of personnel to work on the contractor site, which included military and civilian personnel, as well as Defense Contract Management Agency[7] personnel and system engineering contractors. Greater presence at the contractor's factory has enabled the government to find out about problems as they happen and work with the contractor to come up with solutions and resolve issues quicker, according to GPS program officials.

Nonetheless, the program has experienced more technical problems. For example, last year, during the first phase of thermal vacuum testing (a critical test to determine space-worthiness that subjects the satellite to space-like operating conditions), one transmitter used to send the navigation message to the users failed. The program suspended testing in August 2008 to allow time for the contractor to identify the causes of the problem and take corrective actions. The program also had difficulty maintaining the proper propellant fuel-line temperature; this, in addition to power failures on the satellite, delayed final integration testing. In addition, the satellite's reaction wheels, used for pointing accuracy, were redesigned because on-orbit failures on similar reaction wheels were occurring on other satellite programs—this added about $10 million to the program's cost.

As a result of these problems, the IIF program experienced cost increases and schedule delays. The launch of the first IIF satellite has been delayed until November 2009—almost 3 years late. According to the program office, the cost to complete GPS IIF will be about $1.6 billion—about $870 million over the original cost estimate of $729 million.

In addition, in 2006 we testified[8] that diffuse leadership over military space acquisitions was another factor contributing to late delivery of capability and cost growth. We noted that the diverse array of officials and organizations involved with a space program has made it difficult to pare back and control requirements. GPS was one example we cited. According to the Air Force, in 1998 the government decided to accelerate the fielding of new civil and military GPS signals and added requirements for these signals to the IIR and IIF GPS satellites. These new requirements drove design changes and resulted in technical issues, cost overruns, and program delays.

Problems Experienced in GPS IIF Seen in Other Space System Acquisitions

The problems experienced on the IIF program are not unlike those experienced in other DOD space system acquisitions. We have previously reported that the majority of major acquisition programs in DOD's space portfolio have experienced problems during the past two decades that have driven up costs, caused delays in schedules, and increased technical risk.[9] DOD has restructured several programs in the face of delays and cost growth. At times, cost growth has come close to or exceeded 100 percent, causing DOD to nearly double its investment without realizing a better return on investment. Along with the increases, many programs are experiencing significant schedule delays—as much as 7 years—postponing delivery of promised capabilities to the warfighter. Outcomes have been so disappointing in some cases that DOD has gone back to the drawing board to consider new ways to achieve the same, or less, capability.

Our work has identified a variety of reasons for the cost growth, many of which surfaced in GPS IIF. Generally, we have found that DOD starts its space programs too early, that is before it has assurance that the capabilities it is pursuing can be achieved within resources and time constraints. We have also tied acquisition problems in space to inadequate contracting strategies; contract and program management weaknesses; the loss of technical expertise; capability gaps in the industrial base; tensions between labs that develop technologies for the future and current acquisition programs; divergent needs in users of space systems; and other issues that have been well documented.

We also noted that short tenures for top leadership and program managers within the Air Force and the Office of the Secretary of Defense have lessened the sense of accountability for acquisition problems and further encouraged a view of short-term success.[10] Several other studies have raised similar issues. In 2003, a study[11] conducted for the Defense Science Board, for example, found that government capabilities to lead and manage the space acquisition process have seriously eroded, particularly within program management ranks. A 2005 Defense Science Board study[12] focused specifically on the future of GPS found that the program was hampered by sometimes overlapping, sometimes disconnected roles of Office of the Secretary of Defense staff components, the Joint Staff, and the Air Force. More recently, a commission[13] formed pursuant to the John Warner National Defense Authorization Act for Fiscal Year 2007[14], concluded in 2008 that there is currently no single authority responsible for national security space—which includes GPS—below the President and that within DOD authorities are spread among a variety of organizations, including the Office of the Secretary of Defense, the Air Force, the other military services, the Missile Defense Agency, and the National Reconnaissance Office with no effective mechanism to arrive at a unified budget and set priorities. A study[15] chartered by the House Select Committee on Intelligence also recently found leadership for space acquisitions to be too diffused at higher levels and that there are critical shortages in skilled program managers. While recent studies have made recommendations for strengthening leadership for space acquisitions, no major changes to the leadership structure have been made in recent years. In fact, an "executive agent" position within the Air Force which was designated in 2001 to provide leadership has not been filled since the last executive resigned in 2005.

GPS IIF acquisition problems have not been as extreme as those experienced on other efforts such as the Space Based Infrared System (SBIRS) and the National Polar-orbiting Operational Environmental Satellite System (NPOESS). At the same time, however, the program was not as technically complex or ambitious as these efforts.

DOD Is Implementing Lessons Learned from the GPS IIF Program as It Starts the GPS IIIA Program, but Schedule Is Optimistic

The Air Force is taking measures to prevent the problems experienced on the GPS IIF program from recurring on the GPS IIIA program. However, the Air Force will still be challenged to deliver IIIA on time because the satellite development schedule is compressed. The Air Force is taking the following measures:

- using incremental or block development, where the program would follow an evolutionary path toward meeting needs rather than attempting to satisfy all needs in a single step;
- using military standards for satellite quality;
- conducting multiple design reviews, with the contractor being held to military standards and deliverables during each review;
- exercising more government oversight and interaction with the contractor and spending more time at the contractor's site; and
- using an improved risk management process, where the government is an integral part of the process.

In addition, the Under Secretary of Defense for Acquisition, Technology, and Logistics specified additional guidance for the GPS IIIA program. This includes

- reevaluating the contractor incentive/award fee approach;
- providing a commitment from the Air Force to fully fund GPS IIIA in Program Objectives Memorandum[16] 2010;

- funding and executing recommended mitigation measures to address the next generation operational control segment and the GPS IIIA satellites;
- combining the existing and new ground control segment levels of effort into a single level of effort, giving the Air Force greater flexibility to manage these efforts;
- not allowing the program manager to adjust the GPS IIIA program scope to meet increased or accelerated technical specifications, system requirements, or system performance; and
- conducting an independent technology readiness assessment of the contractor design once the preliminary design review is complete.

Table 2 below highlights the major differences in the framework between the GPS IIF and GPS III programs.

Table 2. Key Differences in Program Framework for GPS IIF and GPS III

	GPS IIF	GPS III
Requirements	Addition of requirements after contract award.	Not allowing an adjustment to the program to meet increased or accelerated requirements.
Development	Immature technologies.	Incremental development, while ensuring technologies are mature.
Oversight	Limited oversight of contractor, relaxed specifications and inspections, and limited design reviews.	More contractor oversight with government presence at contractor facility; use of military standards; and multiple levels of preliminary design reviews, with the contractor being held to military standards and deliverables during each review.

Source: GAO analysis based on discussion with the GPS program office and program documentation.

The Air Force's Schedule for GPS IIIA May Be Optimistic

While these measures should put the GPS IIIA program on sounder footing, the program is facing serious obstacles—primarily in terms of its ability to deliver satellites on schedule. At present, the GPS IIIA program is on schedule and program officials contend that there is no reason to assume that a delay is likely to occur. They point out that the Air Force is implementing an incremental development approach and GPS IIIA, the first increment of GPS III, is not expected to be as technically challenging as other space programs. In addition, program officials point out that the Air Force began risk reduction activities in 1998, and has made a concerted effort to exert more oversight over its contractors and ensure key decisions are backed by sufficient knowledge about technologies, design, and production.

We recognize that these steps offer the best course for GPS to be completed on time. However, we believe there is still considerable risk that the schedule may not be met for the following reasons.

- First, the GPS IIIA program got off to a late start. The program was originally scheduled to begin development in August 2007. However, according to GPS program officials, the Air Force shifted funds from GPS III to other commitments in its space portfolio and to address problems in other programs. The Defense Space Acquisition Board approved formal initiation of the GPS IIIA acquisition in May 2008.
- Second, when compared to other DOD satellite programs, the GPS IIIA program schedule appears highly compressed. The Air Force is planning to launch the first GPS IIIA satellite in 2014 to sustain the GPS constellation. To launch in 2014, the Air Force has scheduled 72 months from contract award to first satellite launch. This schedule is 3 years shorter than the schedule the Air Force has so far achieved under its IIF program. In fact, the time period between contract award and first launch for GPS IIIA is shorter than most other major space programs we have reviewed (see figure 3). Moreover, GPS IIIA is not simply a matter of replicating the IIF program. Though the contractor has had previous experience with GPS, it is likely that the knowledge base will need to be revitalized. The contractor is also being asked to develop a larger satellite bus to accommodate future GPS increments IIIB and IIIC. In addition, the contractor is being asked to increase the power of a new military signal by a factor of 10. In our opinion, there is little room in the schedule to accommodate difficulties the contractor may have in meeting either challenge. In addition, the GPS III program office still has not been able to fill critical contracting and engineering positions needed to assist in satellites design and contractor oversight—both of which functions are to receive more emphasis on this program than in the past. Consequently, the concerns that GPS IIIA could experience a delay are not unreasonable.

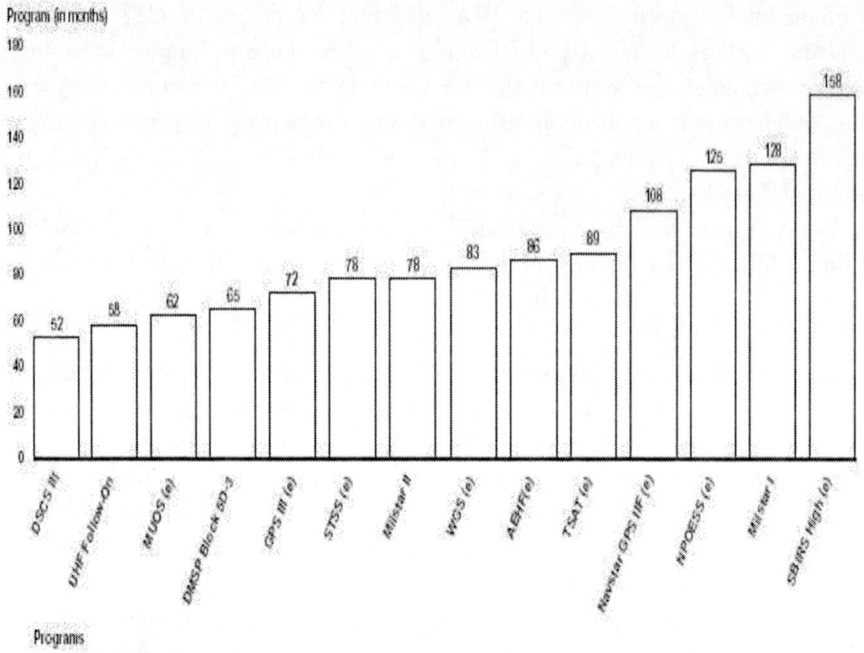

Source: GAO analysis based on program documentation.

Note: DSCS – Defense Satellite Communications System. UHF – Ultra High Frequency. MUOS – Mobile User Objective System. DMSP – Defense Meteorological Satellites Program. GPS – Global Positioning System. STSS – Space Tracking and Surveillance System. WGS – Wideband Global Satellite Communications. AEHF – Advanced Extremely High Frequency. TSAT – Transformational Satellite Communications System. NPOESS – National Polar-orbiting Operational Environmental Satellite System. SBIRS – Space Based Infrared System. All programs with (e) denotation used current estimated dates for launch.

Figure 3. Schedule Development from Start to Launch for Space Programs (in Months)

However, according to DOD officials, the incremental approach to GPS acquisition should significantly lower the risk of schedule delays. Nonetheless, no major satellite program undertaken in the past decade has met its scheduled goals.

- Third, we compared the Air Force's GPS IIIA schedule to best practices associated with effective schedule estimating. Past GAO work has identified nine practices associated with effective schedule estimating. We analyzed the Air Force's GPS IIIA schedule according to these practices and found that one was met, one was not met, and the other seven practices were only partially met. The practices deal with how well the schedule identifies key development activities, the times to complete these activities, as well as the amount of float time associated with each of these activities—float time is the amount of time a task can slip before affecting the critical path. Further, the practices assess how well activities have been integrated with other tasks and whether reserve times have been allocated to high-risk activities. The primary purpose of all scheduling activities is to establish a credible critical path. The best practices have been designed to support that goal. Because the GPS IIIA schedule does not follow all of the best practices, the reliability of the critical path identified in the schedule is diminished.

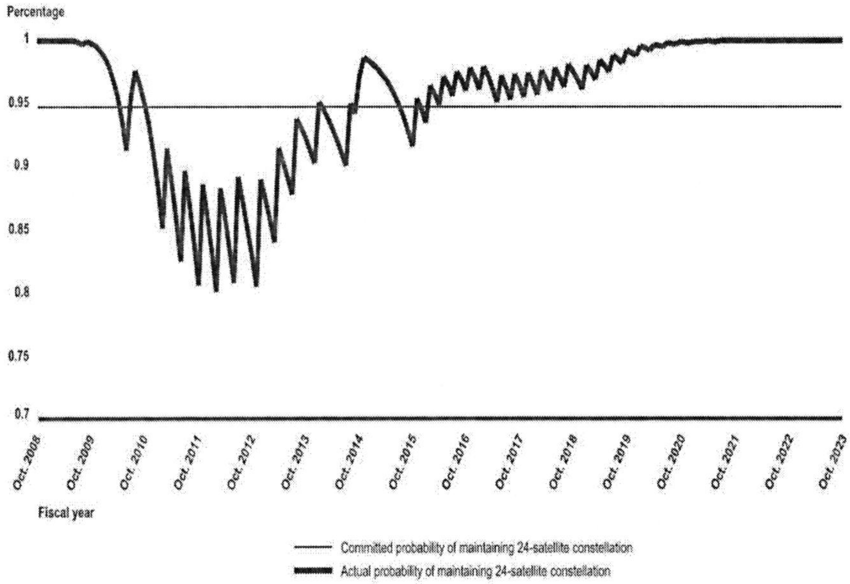

Source: GAO analysis of DOD data.

Figure 4. Probability of Maintaining a Constellation of at Least 24 GPS Satellites Based on Reliability Data and Launch Schedule as of March 2009

A Delay in GPS III Could Severely Impact GPS Users

Delays in the launch of the GPS IIIA satellites will increase the risk that the GPS constellation will decrease in size to a level where it will not meet some users' needs. If the GPS constellation falls below the number of satellites required to provide the level of GPS service that the U.S. government has committed to providing, some military and civilian operations could be affected. DOD is evaluating different approaches that could potentially mitigate the gap. However, procurement of additional GPS IIF satellites does not appear to be feasible.

A Delay in GPS III Could Affect GPS Constellation Performance

The performance standards for both (1) the standard positioning service provided to civil and commercial GPS users and (2) the precise positioning service provided to military GPS users commit the U.S. government to at least a 95 percent probability of maintaining a constellation of 24 operational GPS satellites. Because there are currently 31 operational GPS satellites of various blocks, the near-term probability of maintaining a constellation of at least 24 operational satellites remains well above 95 percent. However, DOD predicts that over the next several years many of the older satellites in the constellation will reach the end of their operational life faster than they will be replenished, and that the constellation will, in all likelihood, decrease in size. Based on the most recent satellite reliability and launch schedule data approved in March 2009, the estimated long-term probability of maintaining a constellation of at least 24 operational satellites falls below 95 percent during fiscal year 2010 and remains below 95 percent until the end of fiscal year 2014, at times falling to about 80 percent. See figure 4 for details.

The probability curve in figure 4 was generated using unique reliability curves for each operational satellite in the current on-orbit GPS constellation, and block-specific reliability curves for each production (unlaunched) GPS satellite, including IIR-M, IIF, IIIA, IIIB, and IIIC satellites. (See app. I for a more complete description of the approach used to generate this probability curve.) Because the reliability curves associated with new blocks of GPS satellites are based solely on engineering and design analysis instead of actual on-orbit performance, this estimated long-term probability of maintaining a 24-satellite constellation could change once actual on-orbit performance data become available. For example, while the block IIA satellites were designed to last only 7.5 years on average, they have actually lasted about twice as long. If GPS IIF satellites were to last twice as long as their currently estimated mean life expectancy of 11.5 years, the probability of maintaining a larger constellation would increase, but the long-term probability of maintaining the 24-satellite constellation would not improve significantly. Moreover, program officials provided no evidence to suggest that the current mean life expectancy for IIF satellites is overly conservative.

A delay in the production and launch of GPS III satellites could severely impact the U.S. government's **ability to meet its commitment to maintain a 24-**satellite GPS constellation. The severity of the impact would depend upon the length of the delay. For example, a 2-year delay in the production and launch of the first and all subsequent GPS III satellites would reduce the probability of maintaining a 24-satellite constellation to about 10 percent by around fiscal year 2018. This significant gap in service would persist for about 2 years before the constellation began to recover. Moreover, this recovery—that is, the return to a high probability of maintaining a 24-satellite constellation—would take an additional 2 to 3 years. Consequently, a 2-year delay in the production and launch of GPS III satellites would most likely result in a period of roughly 5 years when the U.S. government would be operating a GPS constellation of fewer than 24 satellites, and a 12-year period during which the government would not meet its commitment to maintaining a constellation of 24 operational GPS satellites with a probability of 95 percent or better. For example, the delay in GPS III would reduce the probability of maintaining a 21-satellite constellation to between 50 and 80 percent for the period from fiscal year 2018 through fiscal year 2020. Moreover, while the probability of maintaining an 18-satellite constellation would remain relatively high, it would still fall below 95 percent for about a year over this period. See figure 5 for details.

Both Military and Civilian GPS Users Would Be Affected by a Delay in GPS III

The impacts to both military and civil users of a smaller constellation are difficult to precisely predict. For example, a nominal 24-satellite constellation with 21 of its satellites broadcasting a healthy standard positioning service signal would continue to satisfy the availability standard for good user-to-constellation geometry articulated in the standard positioning service performance standard.[17] However, because the GPS constellation has been operating above the committed performance standard for so long, military and civil users have come to expect a higher level of service, even though this service is not committed to them. Consequently, some users may sense an operational impact even if the constellation were to perform at or near its committed standards. In general, users with more demanding requirements for precise location solutions will likely be more impacted than other users.[18] During our interviews with military, civil, and commercial representatives, several examples of possible impacts of a smaller GPS constellation were discussed.

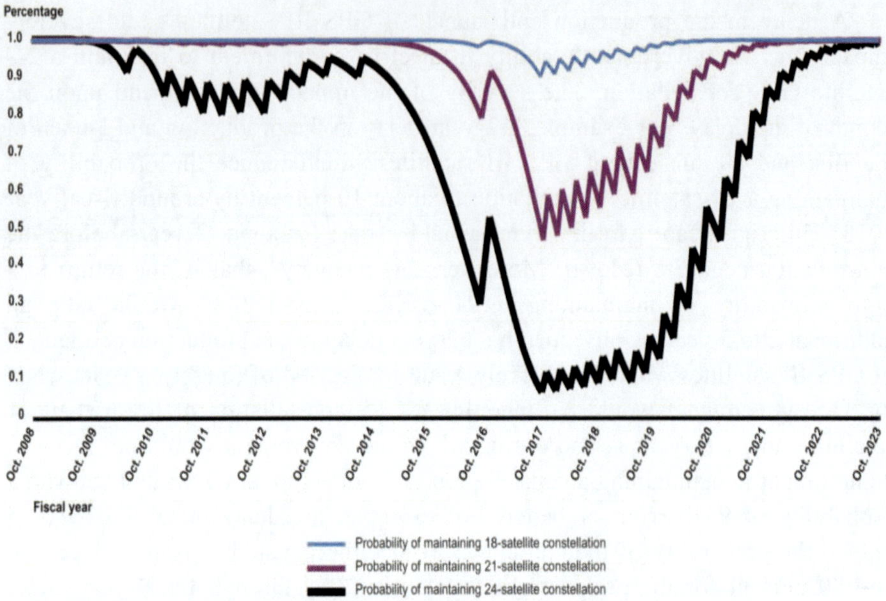

Source: GAO analysis of DOD data.

Figure 5. Probability of Maintaining a Constellation of at Least 18, 21, and 24 GPS Satellites Based on Reliability Data as of March 2009 and a 2-Year GPS III Launch Delay

- The accuracy of precision-guided munitions that rely upon GPS to strike their targets could decrease. To accomplish their mission, military forces would either need to use larger munitions or use more munitions on the same target to achieve the same level of mission success. The risks of collateral damage could also increase.
- Intercontinental commercial flights use predicted satellite geometry over their planned navigation route, and may have to delay, cancel or reroute flights.
- Enhanced-911 services, which rely upon GPS to precisely locate callers, could lose accuracy, particularly when operating in “urban canyons” or mountainous terrain.

Another important consideration is that both the standard positioning service and precise positioning service performance standards assume that users have unobstructed visibility to nearly the entire sky,[19] an assumption that does not hold for the large number of users operating in moderately mountainous terrain, in the “urban canyons” of large cities, or under forest foliage.

Different Approaches Are Being Evaluated that Could Potentially Mitigate the Gap

The Air Force is aware that there is some risk that the number of satellites in the GPS constellation could fall below its required 24 satellites, and that this risk would grow significantly if the development and launch of GPS IIIA satellites were delayed. Consequently, an Air Force Space Command representative informed us that the command has established an independent review team to examine the risks and consequences of a smaller constellation on military and civil users. However, at this time, Air Force representatives believe that the best approach to mitigating the risk is to take all reasonable steps to ensure that the current schedule for GPS IIF and III is maintained. Those steps include a commitment from the Air Force to fully fund GPS IIIA in the fiscal year 2010 Program Objectives Memorandum, and use of an incremental development approach toward meeting needs. This incremental approach would place a premium on controlling schedule risk by, among other things, deferring consideration of civil requirements for subsystems like the Distress Alerting Satellite System (DASS) and the Satellite Laser Ranging (SLR) payloads to GPS IIIB or GPS IIIC satellite blocks.

Options for developing lower-cost alternatives to current GPS satellites appear to be very limited. For example, in 2007 the Air Force Scientific Advisory Board examined whether small satellites—which can be developed more quickly and at relatively low cost—might help meet some PNT mission requirements. The board concluded that small satellites may eventually have operational utility in augmenting GPS III capabilities, with emphasis on enhancing the utility of the GPS M-code signal's capabilities against jamming. However, the need for an extensive control segment infrastructure to monitor and control these small satellite augmentations, combined with the need to develop, produce, and install user equipment, would make it very challenging to field a near-term small satellite augmentation for PNT. With respect to providing basic PNT services, the board noted that studies of PNT satellite constellations, performed at different times and by different organizations in the United States and elsewhere, demonstrate that a robust constellation of relatively powerful satellites operating at medium earth orbit is the best way to provide continuous worldwide PNT services; this is a performance set that small satellites currently cannot provide.

According to Air Force representatives, the procurement of additional IIF satellites is not feasible, and initiating development of an alternative full- scale, satellite-based PNT system appears to be impractical. Such a system would likely be very expensive and would compete with GPS III development for funding, making it harder for the Air Force to meet its commitment to fully fund GPS IIIA development. Moreover, the GPS III system development contract was awarded in accordance with an approved GPS III acquisition strategy, which selected one alternative from two competing contractors' designs; an alternative system development would be, in effect, a significant deviation from that approved strategy. Finally, it seems unlikely that the award of a separate system development contract with another contractor would have any real impact on reducing the risk of delivering GPS IIIA requirements on the current schedule.

In the event that this strategy proves unsuccessful and the schedule for GPS III slips, additional measures could be considered. For example, excluding random failures, the operational life of a GPS satellite tends to be limited by the amount of power that its solar arrays can produce. This power level declines over time as the solar arrays degrade in the space environment until eventually they cannot produce enough power to maintain all of the satellite's subsystems. However, according to Air Force representatives, the effects of this power loss can be mitigated somewhat by actively managing satellite subsystems—shutting them down when not needed—thereby reducing the satellite's overall consumption of power. It would also be possible to significantly reduce the satellite's consumption of power by shutting off a secondary GPS payload. This would buy additional time for the navigation mission of the satellite at the expense of the mission supported by the secondary payload. The 2004 U.S. Space- Based Positioning, Navigation and Timing (PNT) policy affirmed PNT as the primary mission for the GPS constellation, and stated that no secondary payload may adversely affect the performance, schedule, or cost of GPS, its signals, or services. Nevertheless, at this time the Air Force has no intention of shutting off the secondary GPS payload. Moreover, until there is a more immediate risk that the constellation will fall below its required size, there is no reason to take this step.

Military and civil users might also take steps in response to a smaller GPS constellation. While a smaller GPS constellation could result in a significant reduction in positioning and navigation accuracy at certain times and locations, these times and locations are usually predictable in near-real time. Consequently for military users, who must rely upon GPS's precise positioning service, a smaller constellation could require changes in its approach to mission planning to ensure that operations are conducted at times when GPS accuracy is relatively high, or changes in tactics employed during a mission. For example, military users could utilize a larger number of (or more powerful) munitions to achieve an equivalent level of mission effectiveness.

For civil and commercial users, one possible impact of a smaller GPS constellation could be an increased use of other positioning, navigation, and timing services, including those expected to be offered through Europe's Galileo system by the middle of the next decade. U.S. government officials at the various civil agencies and departments clearly understand what the government has committed to through GPS and they all have designed programs to function with this limit, with augmentations.

New Satellite Capabilities Will Not Be Leveraged Because of Delayed Delivery of Ground and User Equipment Capabilities

To maximize the benefit of GPS, the deployment of its space, ground control, and user equipment capabilities must be synchronized so that the full spectrum of military assets—weapons, aircraft, and ships, for example—and individual users can take advantage of new capabilities such as added protection from jamming. However, because of funding shifts and diffuse leadership, the Air Force has not been successful in synchronizing space, ground control, and user equipment segments. As a result of the poor synchronization, new GPS capabilities may be delivered in space for years before military users can take advantage of these capabilities.

Air Force Has Deferred the Delivery of Ground Control Capabilities

The Air Force used funding set aside for the ground control segment to resolve GPS IIF development problems, causing a delay in the delivery of new ground control capabilities. The GPS ground control segment has evolved over time from the Operational Control Segment (OCS) to the current Architecture Evolution Plan (AEP). GPS IIIA satellites are to be controlled by a future ground control system called Next Generation Control Segment, or OCX. OCS was supposed to control and exploit GPS IIF space capabilities. However, because of the addition of new requirements and technical issues on the IIF program, funding was diverted from OCS to GPS IIF satellite development efforts. As a result, the delivery of new ground control capabilities will occur later than originally planned.

Table 3 illustrates satellite functions and capabilities that have yet to be made operational through the ground control segment.

Table 3. Delays in Delivery of GPS Operational Functionality

Function or capability enabled	Original ground control program/version	Current or future ground control program/ version	Amount of delay (in months)
GPS IIR-M satellites (first launch in 2005 & currently being launched)			
Command & telemetry for IIA & IIR and satellites, and use of additional signals	OCS Version 5.0 September 2005	OCS Version 5.2.1 September 2007	24
Command & telemetry for IIRM & IIF satellites	OCS Version 5.0 September 2005	AEP Version 5.2.2 March 2008	30
Selective Availability Anti-Spoofing Module	OCS Version 5.0 September 2005	AEP Version 5.5 September 2009	48
Second civil signal (L2C)	OCS Version 6 September 2007	OCX Block I or II September 2012/ September 2013	60-72
Military code (M-code)	OCS Version 6 September 2007	OCX Block I or II September 2012/ September 2013	60-72
GPS IIF satellites (first launch planned for November 2009)			
Third civil signal (L5)	OCS Version 6 September 2007	OCX Block I or II September 2012/ September	60-72

Source: GPS program office.

For example, in 2005 the Air Force began launching its GPS IIR-M satellites that broadcast a second civil signal (the L2C). Unfortunately, the ground control segment will not be able to make the second civil signal operational until late 2012 or 2013.

By delaying the delivery of ground control capabilities, the Air Force has created an imbalance between the capabilities offered by GPS satellites and the ability to exploit and make operational these capabilities through the ground control segment.

Synchronization Problems Will also Delay Fielding of Improved GPS Capabilities to Military Users

GPS satellites that will broadcast the modernized military signal require military user equipment capable of receiving and processing the signal so that military users can take advantage of the improved military capabilities.

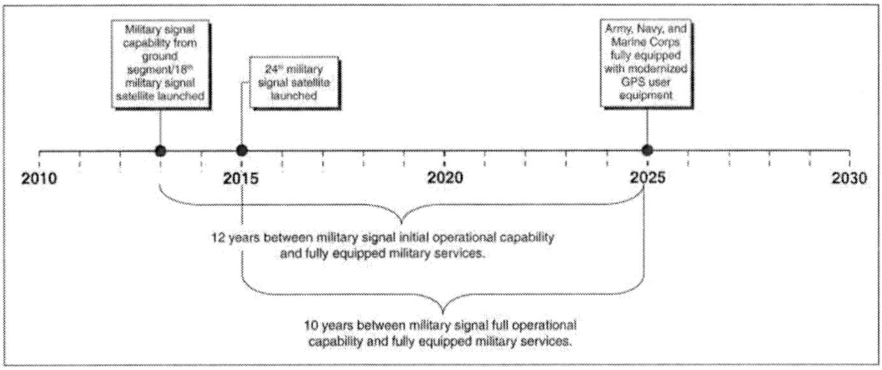

Source: GAO analysis of DOD documents and discussions with DOD officials.

Figure 6. Gap in the Ability of the Military to Use the Modernized Signal

Before the modernized military signal can be considered initially operational, it must be broadcast from at least 18 satellites, which is expected to occur in 2013. For full operational capability, it must be broadcast from 24 satellites, which is expected to occur in 2015.Consequently, the new military signal will be made operational by the GPS satellites and ground control system in about 2013, but the warfighter will not be able to take full advantage of this new signal until about 2025—when the modernized user equipment is completely fielded. See figure 6 for our analysis of the gap between when the modernized military signal will be available on the GPS satellites and when the military services will be able to take advantage of it.

Funding and Technical Issues Have Delayed User Equipment Development, but the Air Force Is Seeking to Accelerate Development

The Air Force will spend the next several years developing prototype cards and production-ready receiver hardware for selected platforms within the space, air, ground, and maritime environments. Even after this is done, the services will still need to add the new user equipment to other platforms, which could take 10 or more years. This is due to the fact that the integration and installation of the new user equipment on the remaining platforms has to be coordinated with existing upgrade schedules for those platforms. As a result, the services' ability to achieve a joint military navigation warfare capability, an essential element in conducting future military operations, may not be realized until 2025 based on user equipment delivery schedules.

Funding issues are a contributing factor in the delay in fielding new user equipment. According to Air Force officials, the GPS program office focused on developing the satellites, particularly when technical problems arose. Funding was diverted from the user equipment program to the GPS satellite program to fix problems, which resulted in delays in the development and acquisition of the user equipment.

Diffused leadership has been particularly problematic in terms of DOD's ability to synchronize delivery of space, ground control, and user equipment assets. The responsibility for developing and acquiring GPS satellite and associated ground control segments and for acquiring and producing user equipment for selected platforms for space, air, ground, and maritime environments falls under the Air Force's Space and Missile Systems Center. On the other hand, responsibility for acquiring and producing user equipment for all other platforms falls on the military services. Figure 7 illustrates how the responsibilities for developing, acquiring, and producing GPS user equipment are divided among the services.

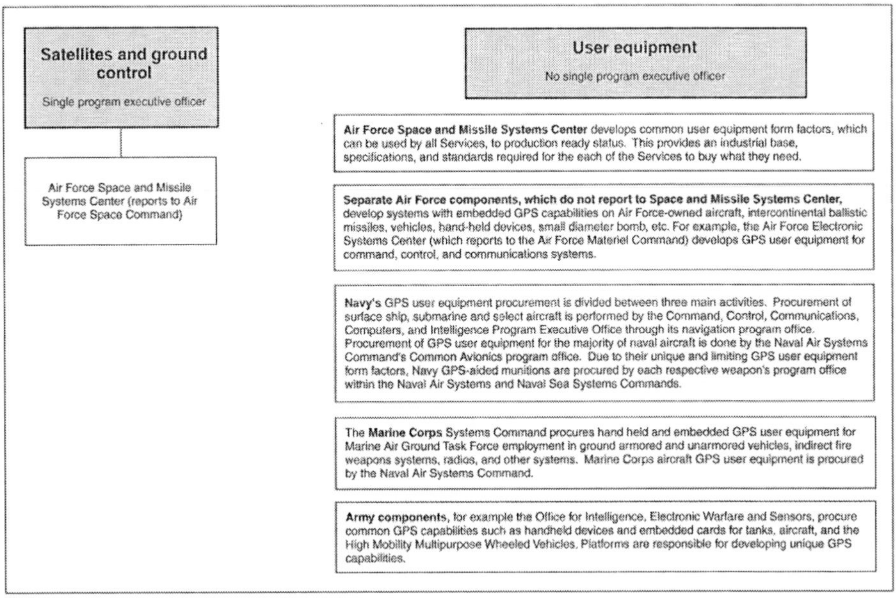

Source: GAO presentation of DOD data.

Figure 7. Responsibilities Among the Military Services for Procurement of GPS User Equipment

Because different military services are involved in developing user equipment for the weapon systems they own and operate, there are separate budget, management, oversight, and leadership structures over the space and ground control and the user equipment segments. As such, there is no single authority responsible for synchronizing all the procurements and fielding related to GPS. A 2008 U.S. Strategic Command Functional Solutions Analysis, conducted to provide recommendations for solutions to positioning, navigation, and timing gaps, noted that the Air Force is responsible for developing and integrating military GPS user equipment for select platforms, and that integration and testing of these platforms is required to be complete so that the user equipment is available for procurement when the military signal becomes operational. However, this analysis showed no military service program office commitment of resources for procuring military GPS user equipment in service programming documents. Furthermore, DOD's management attention has been focused on delivering space capabilities. Only recently has DOD begun to shift its focus by recognizing that the user equipment segment needs to play an equal role in the overall GPS synchronization effort.

Efforts to Speed up Delivery of User Equipment Face Obstacles

There have been various recommendations to accelerate the fielding of modernized military user equipment, though there are obstacles in the way of implementation. In October 2005, the Defense Science Board[20] recommended that DOD initiate an aggressive program to introduce antijam enhancements as soon as possible. In August 2006, OSD issued a GPS User Equipment Development and Procurement Policy, which mandated that certain equipment categories have the modernized GPS user equipment by the time the 24th military code satellite is declared operational. In June 2007, representatives from the Combatant Commands, U.S. Strategic Command, and U.S. Joint Forces Command requested that an aggressive schedule be established for all GPS segments to achieve military code initial operational capability by fiscal year 2013. In March 2008, the Joint Requirements Oversight Council recommended that the Air Force adjust the development and acquisition of the modernized GPS user equipment to ensure that warfighters can use space-based capabilities. Recommendations included amending programmatic schedules and funding profiles to incorporate military code capabilities at or before the initial operational capability date.

To accelerate the delivery of the new user equipment, the Air Force increased the user equipment budget by $272 million for fiscal years 2009 through 2011. In the conference reports accompanying the Department of Defense Appropriation Act for Fiscal Year 2008 and the National Defense Authorization Act for Fiscal Year 2008, conferees recommended an additional $63.2 million in funding for GPS user equipment. However, the additional funds will not speed up development of the new user equipment to a large extent, because the program office is experiencing technical issues in developing the prototype cards. The major technical issue is with the difficulty in moving to a new security architecture, Protection of Navigation, which will provide information assurance.

According to a GPS program office official, OSD, the Air Staff, U.S. Strategic Command, Air Force Space Command, and the GPS program office are looking at ways to get some of the modernized military user equipment to the field sooner. However, there are challenges with this approach, particularly because certain security requirements—antispoof,[21] antijam, and antitamper—should be met before user equipment can be fielded in conflict situations. According to an official at the GPS program office, meeting these security requirements is proving to be technically challenging, and attempting this at an accelerated rate is risky.

Prudent Steps Taken so GPS Can Meet Broader Needs but Challenges Exist in Coordinating Requirements and Ensuring Compatibility

GPS has produced dramatic economic and security improvements both for the United States and globally. Ensuring that it can continue to do so is extremely challenging given competing interests, the span of government and commercial organizations involved with GPS, and the criticality of GPS to national and homeland security and the economy. On the one hand, DOD must ensure military requirements receive top priority and the program stays executable. In doing so, it must ensure that the program is not encumbered by requirements that could disrupt development, design, and production of satellites. On the other hand, there are clearly other enhancements that could be made to GPS satellites that could serve a variety of vital missions—particularly because of the coverage GPS satellites provide—and there is an expressed desire for GPS to serve as the world's preeminent positioning, navigation, and timing system. In addition, while the United States is challenged to deliver GPS on a tight schedule, other countries are designing and developing systems that provide the same or enhanced capabilities. Ensuring that these capabilities can be leveraged without compromising national security or the preeminence of GPS is also a delicate balancing act that requires close cooperation between DOD, the Department of State, and other institutions.

Because of the scale and number of organizations involved in maximizing GPS, we did not undertake a full-scale review of requirements and coordination processes. However, we reviewed documents supporting these processes and interviewed a variety of officials to obtain views on its effectiveness. While there is a consensus that DOD and other federal organizations involved with GPS have taken prudent steps to manage requirements and optimize GPS use, we also identified challenges in the areas of ensuring civilian requirements can be met and ensuring that GPS is compatible with other new, potentially competing global space-based positioning, navigation, and timing systems.

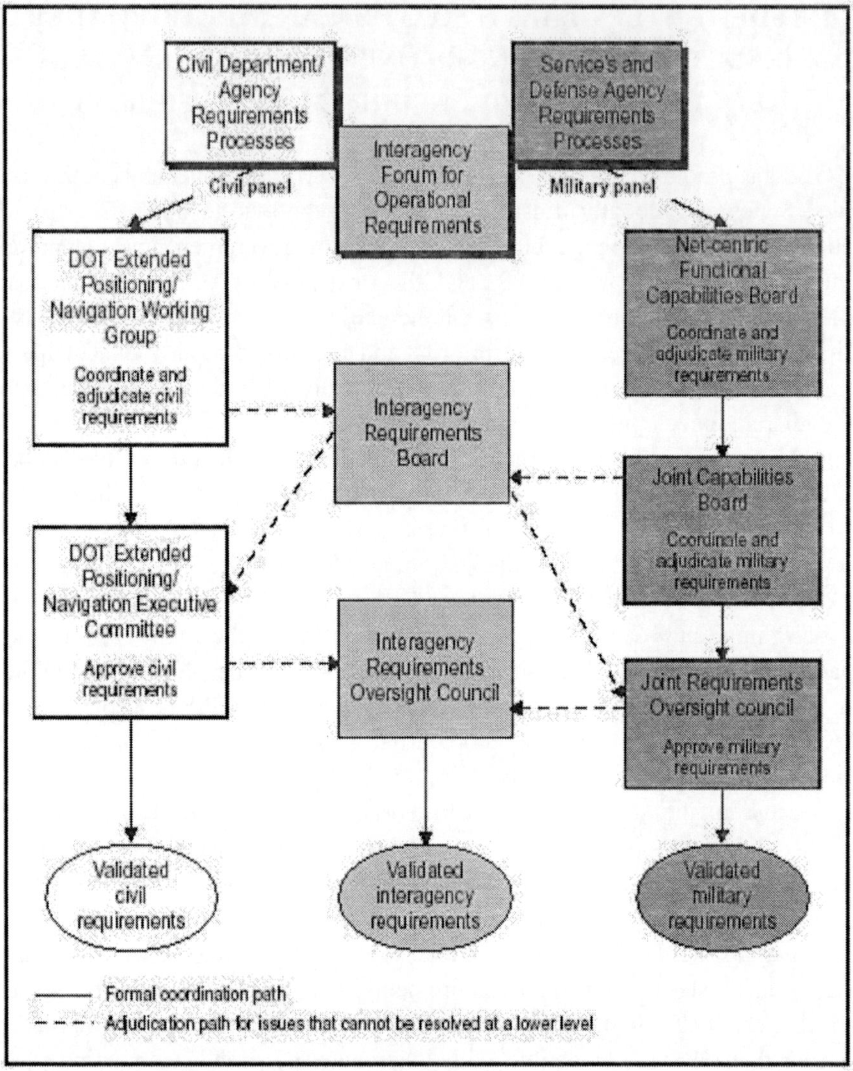

Source: GAO presentation of Department of Transportation and U.S. Air Force data.

Figure 8. Interagency Process for Submitting and Validating GPS Requirements

The Process for Approving GPS Civil Requirements is Rigorous but Untested

The 2004 U.S. Space-Based Positioning, Navigation and Timing (PNT) policy provides guidance for civil involvement in the development of requirements for the modernization of GPS capabilities and the requirements process includes an entry point for civil requirements. This entry point is the Interagency Forum for Operational Requirements (IFOR), working groups consisting of a civil and a military panel. The IFOR receives proposed GPS requirements from civil agencies and assists in developing and validating them. From this point, the proposed requirement follows a DOD and civil path to validation with involvement from various interagency boards and councils. Figure 8 illustrates this formal process for submitting, considering, and validating civil GPS requirements.

However, according to civil agencies, the analyses and documentation called for under the process are confusing and time-consuming.

While GPS remains critical to national security and military operations, government policy calls for GPS planning to consider integration of civil requirements for the civilian infrastructure. The process for considering civil GPS requirements is intended to maintain fiscal discipline by ensuring only critical needs are funded and developed. Specifically, the process requires that civil agencies internally identify and validate their proposed requirements, and conduct cost, risk, and performance analyses. Our past work has shown that requirements add-ons are a major source of acquisition instability. In this case, the formal process also requires that the agency proposing the requirement pay the costs associated with adding it to the GPS III satellites, thereby forcing agencies to separate their wants from needs.

Civil Agencies Find the GPS Requirements Process Confusing

According to the civil agencies that have proposed GPS requirements, the formal requirements approval process is confusing and time-consuming. Specifically, they stated that DOD's documentary and analysis standards are new to civil agencies and therefore difficult and time-consuming for them to manage. Some agencies have reported that it is costly for them to pay for the more detailed supporting analyses requested by DOD. For example, one civil agency had to withdraw and resubmit a proposal for new GPS requirements because it lacked necessary information, including a cost-benefit analysis. Furthermore, civil agencies' submitted requirements have necessitated that DOD perform further studies on compatibility and integration issues to ensure that the proposed requirements will not adversely affect the primary GPS mission.

The two civil requirements that have entered the requirements process are the Distress Alerting Satellite System (DASS) and the geodetic[22] requirement implemented by Satellite Laser Ranging (SLR). Both are joint civil and military mission requirements and would be potential secondary payloads on GPS. DASS is an electronic unit that will receive beacon signals identifying a distressed individual's location and transmit this location data to emergency responders. The SLR laser retroreflector, which weighs less than 7 pounds, is being considered for inclusion starting with increment IIIB satellites. Scientists would aim a laser to the reflector to more precisely determine the satellite's position, ultimately allowing for more precise measurements on the ground. This SLR capability would support users who need to make very accurate measurements for scientific applications.

- Distress Alerting Satellite System: The Coast Guard submitted the DASS requirement to the IFOR in 2003. Early in the review process, a debate on whether DASS was a civil or military requirement ensued. The IFOR decided to have military and civil panels review the requirement and resubmit it through the Joint Capabilities Integration and Development System (JCIDS) process. It took a total of 5 years to resolve the debate and prepare and resubmit the package. In July 2008, the civil agencies submitted DASS requirements and an analysis of alternatives to the IFOR for review. To date, a decision has not yet been made as to if and when the capability will be inserted on GPS satellites.
- Satellite Laser Ranging: In April 2007, NASA submitted the SLR requirements package along with an analysis of alternatives to the IFOR. The IFOR officially accepted the SLR package into the IFOR process in August of that year. However, in June 2008, DOD opposed implement tation of the SLR capability due to integration and compatibility concerns with the GPS satellites. A joint Air Force and NASA working group was established to resolve the integration and compatibility issues and report back to the IFOR by June 2009 prior to moving the requirement from the IFOR into the JCIDS process.

DASS supporters have stated that the GPS constellation is the ideal platform for search and rescue capabilities. The current search and rescue capability is expected to degrade by 2017 and completely fail by 2020. More urgently, supporters say that the Canadian government's offer to provide DASS hardware at a $90 million cost savings to the United States must be acted upon by August

2009 or Canada may provide this component to a developing foreign satellite navigation system. The SLR capability, until recently, existed on two GPS satellites. One satellite was decommissioned, and hence according to NASA does not meet its or other civil agencies' needs to perform scientific and geodetic applications. According to NASA, the SLR would need to be implemented on most of the GPS constellation to meet geodetic requirements for science and other user requirements. If the DOD does not include DASS and SLR on GPS satellites, U.S. users of these capabilities may be dependent on foreign systems which already include, or have plans to include, both DASS-like and SLR capabilities in their satellite navigation systems.

Coordinating GPS Activities with the International Community also Presents Challenges

The U.S. government—specifically the State Department—is faced with challenges in ensuring GPS is compatible and interoperable with other new, potentially competing global space-based positioning, navigation, and timing systems. While the U.S. government has engaged a number of other countries and international organizations in cooperative discussions, only one legally binding agreement has been established. Furthermore, some U.S. manufacturers of GPS receivers stated that European Union manufacturers may have a competitive advantage over U.S. companies with respect to the manufacture and sale of Galileo-capable receivers, though officials with the European Commission disagree. In addition, Department of State officials have expressed concerns over the limited number of technical experts available to support activities under these cooperative arrangements. Without these resources, officials are concerned that it may be difficult to continue to ensure the compatibility and interoperability of foreign systems.

Joint Statements of Cooperation Made and One Agreement Established

The United States has made joint statements of cooperation with Australia, India, Japan, and Russia to promote compatibility and interoperability and mutual interests regarding the civil use of GPS and its augmentations[23] and established an executive agreement with the European Community (see table 4 for a list of types of cooperative arrangements with other countries).[24] The joint statements and executive agreement were sought to avoid interference with each others' systems, and to facilitate the pursuit of common civil signals. Under the national space-based PNT policy, it is the Department of State's role to promote the civil aspects of GPS and its augmentation services and standards with foreign governments and other international organizations. The Department of State leads negotiations with foreign governments and international organizations regarding civil and, as appropriate, military space-based PNT matters including, but not limited to, coordinating interagency review of international agreements with foreign governments and international organizations regarding the planning, operation, management, and or use of the GPS and its augmentations. While most of the cooperative arrangements are joint statements that express the parties' intent to cooperate on GPS-related activities, the United States and the European Commission have established an executive agreement that is considered binding under international law.

Table 4. U.S. Cooperation with Foreign Entities on Satellite Navigation

Country	Cooperative arrangement/effective dates			
	Executive agreement	Joint statement	No agreement	Date signed
Japan		√		1998
EU	√			2004
Russia		√		2004
Australia		√		2007
India		√		2007
China			√	na

Source: GAO analysis of U.S. Department of State data.

U.S. and European Commission Working to Address Concerns Regarding Access to Galileo Information

According to the executive agreement with the European Community, subject to applicable export controls, the United States and the European Commission are to make sufficient information concerning their respective civil satellite-based signals and augmentations publicly available on a nondiscriminatory basis, to ensure equal opportunity for persons who seek to use these signals, manufacture equipment to use these signals, or provide value-added services which use these signals.

In 2006, the European Commission publicly released draft technical specifications for its open service. The draft document requests manufacturers to obtain a commercial license from the European Commission to sell and import products designed to work with the European satellite navigation system, Galileo.

While this licensing requirement applies to all manufacturers, some U.S. companies stated that some foreign user equipment manufacturers who are members of the Galileo consortia may have an unfair advantage over U.S. companies. This is because the Galileo consortia currently have access to testing hardware and may be able to introduce their products more quickly into the marketplace once they are granted a commercial license.

Officials with the European Commission told us that they do not believe the license restrictions or the knowledge gained from testing the Galileo systems are discriminatory. They further stated that the restrictions in obtaining a commercial license to sell user equipment apply to all companies, not just U.S. companies and they have not yet issued licenses to any company. In the meantime, a U.S. and European Commission working group on trade and civil applications is discussing the licensing issue.

However, U.S. firms have raised concerns to the Department of Commerce (Commerce) on the lack of information from the European Commission relating to the process for obtaining a license to sell Galileo equipment. According to Commerce, U.S. firms have asserted that they are not aware of how, where, or when to apply for such a license, despite repeated inquiries to the U.S.-European Commission trade working group and direct contacts with European Commission officials—and the timeline for the licensing process is unknown. Commerce further noted that U.S. manufacturers wanting to enter the Galileo market are hesitant to invest in technology that is not officially licensed and that could possibly be banned from sale. It takes industry 18 to 24 months to develop a market-ready receiver, and the first operational Galileo satellite is scheduled for launch in 2010. U.S. firms are concerned they will not have their products ready by that time and will lose their market share to European companies with inside access to technology and/or licensing information.

State Officials Believe International Efforts Lack Dedicated Resources

According to Department of State officials, the department lacks dedicated technical expertise to monitor international activities. The Department of State relies on a small pool of experts from DOD and the seven civil agencies represented on the National Executive Committee for Space-Based PNT. These experts are often in high demand because they work on other GPS-related activities and in some cases have other assigned duties that are unrelated to GPS. According to the Department of State, in many cases these experts and those in other agencies must continually justify to their managers that their attendance at international meetings is important. Given the progress made in working with foreign governments to establish arrangements, share information, and ensure compatibility and interoperability with GPS, Department of State officials would like DOD and civil agencies to dedicate funding and staff positions to international activities accompanied by a sustained level of senior management support and understanding of the importance of these activities. Without an expanded pool of technical expertise and related resources, Department of State officials stated they are concerned that ongoing international efforts to ensure compatibility of foreign systems with GPS could be jeopardized.

CONCLUSIONS

GPS has enabled transformations in military, civil, other government, and commercial operations and has become part of the critical infrastructure serving national and international communities. Clearly, the United States cannot afford to see its GPS capabilities decrease below its requirement, and optimally, it would stay preeminent. Over the past decade, however, the program has experienced cost increases and schedule delays. While the Air Force is making a concerted effort to address acquisition problems, there is still considerable risk that satellites will not be delivered on time, leading to gaps in capability. Focused attention and oversight are needed to ensure the program stays on track and is adequately resourced, that unanticipated problems are quickly discovered and resolved, and that all communities involved with GPS are aware of and positioned to address potential gaps in service. But this is difficult to achieve given diffuse responsibility over various aspects of the GPS acquisition program. Moreover, disconnects between the space, ground control, and user equipment components have significantly lessened the military's ability to take advantage of enhancements, particularly as they relate to assuring the continuity of service during military engagements. Without more concentrated leadership attention, such disconnects could worsen, particularly since (1) both the ground control and user equipment programs have been subject to funding shifts to pay for problems affecting the satellite segment, and (2) user equipment programs are executed by separate entities over which no one single person has authority. Lastly, ensuring that GPS can continue to produce dramatic improvements to civil agencies' applications, calls for any weaknesses that are identified in the civil agency GPS requirements process to be addressed.

Recommendations for Executive Action

Because of the criticality of the GPS system and potential delays, and given the importance of GPS to the civil community, we are making the following recommendations.

- We recommend that the Secretary of Defense appoint a single authority to oversee the development of the GPS system, including DOD space, ground control, and user equipment assets, to ensure that the program is well executed and resourced and that potential disruptions are minimized. The appointee should have authority to ensure DOD space, ground control, and user equipment are synchronized to the maximum extent practicable; and coordinate with the existing positioning, navigation, and timing infrastructure to assess and minimize potential service disruptions should the satellite constellation decrease in size for an extended period of time.
- We recommend that the Secretaries of Defense and Transportation, as the co-chairs of the National Executive Committee for Space-Based Positioning, Navigation and Timing, address, if weaknesses are found, civil agency concerns for developing requirements, and determine mechanisms for improving collaboration and decision making and strengthening civil agency participation.

Agency Comments and Our Evaluation

DOD concurred with our first recommendation to appoint a single authority to oversee the development of the GPS system, including space, ground control, and user equipment assets, to ensure that the program is well executed, resourced, and that potential disruptions are minimized. DOD stated that it has recognized the importance of centralizing authority to oversee the continuing synchronized evolution of the GPS. According to DOD, the Deputy Secretary of Defense has reaffirmed that the Assistant Secretary of Defense for Networks and Information Integration (ASD NII)) is designated with authority and responsibility for all aspects of the GPS. DOD further stated that the U.S. Air Force is the single acquisition agent with responsibility for synchronized modernization of GPS space, ground control, and military user equipment.

In concurring with our recommendation on appointing a single authority to oversee the development of the GPS system, DOD asserts that ASD NII is designated with authority and responsibility for all aspects of GPS, and that the Air Force is the single acquisition agent responsible for synchronizing GPS segments. In addition, responsibility for acquiring GPS military user equipment acquisitions falls under various officials within the military services. We agree that given the diversity of platforms and equipment variations involved, it would not be realistic for the Air Force to unilaterally produce a "one-size-fits-all" solution. However, this does not obviate the need for a single authority to oversee the development of all GPS military user equipment to better ensure greater coordination with deployed satellite capabilities. Without an approach that enables a single individual to make resource decisions and maintain visibility over progress, DOD is at risk of facing the same issues in synchronizing the delivery of GPS assets and wasting capability that will be available in space but not on the ground. In addition, DOD may still want to consider establishing a means by which progress in developing the satellites and ground equipment receives attention from the highest levels of leadership that is the Secretary and perhaps the National Security Council, given the criticality of GPS to the warfighter and the nation, and the risks associated with not meeting schedule goals.

DOD concurred with our second recommendation to address, if weaknesses are found, civil agency concerns for developing requirements and determine mechanisms for improving collaboration and decision making, and strengthening civil agency participation. DOD acknowledged that it employs a rigorous requirements process and is aware of the frustration civil agencies face when using this process. DOD further indicated that it worked to put in place an interagency requirements plan, and is currently in the process of jointly coordinating the Charter for an Interagency Forum for Operational Requirements to provide venues to identify, discuss, and validate civil or dual-use GPS requirements. Finally, DOD noted that it will continue to seek ways to improve civil agency understanding of the DOD requirements process and work to strengthen civil agency participation. We support DOD's efforts to inform and educate other civil agencies on the requirements process. As it undertakes these efforts, DOD should ensure that it is taking a more active role in directly communicating with civil agencies to more precisely identify concerns or weaknesses in the requirements process.

The full text of DOD's comments may be found in appendix IV. We also received technical comments from the other departments and NASA, which we incorporated where appropriate.

As agreed with your offices, unless you publicly announce the contents of this report earlier, we plan no further distribution of it until 8 days from the report date. At that time, we will send copies of this report to the Secretaries of Defense, Agriculture, Commerce, Homeland Security, Interior, State, and Transportation; the National Aeronautics and Space Administration; and interested congressional committees. The report will also be available at no charge on the GAO Web site at http://www.gao.gov.

If you have any questions about this report or need additional information, please contact me at (202) 512-4841 or chaplainc@gao.gov. Contact points for our Offices of Congressional Relations and Public Affairs may be found on the last page of this report. The major contributors are listed in appendix V.

Cristina T. Chaplain
Director
Acquisition and Sourcing Management

APPENDIX I: SCOPE AND METHODOLOGY

To assess the Global Positioning System (GPS) satellite, ground control, and user equipment acquisition programs and determine whether GPS capabilities are being synchronized, we reviewed and analyzed program plans and documentation related to cost, schedule, requirements, program direction, and satellite constellation sustainment, and compared programmatic data to GAO's criteria compiled over the last 12 years for best practices in system development. We also interviewed officials from Air Force Space and Missile Systems Center GPS program office; Air Force Space Command; Office of the Joint Chiefs of Staff; Office of the Undersecretary of Defense for Acquisition, Technology, and Logistics; Assistant Secretary of Defense Office of Networks and Information Integration; United States Strategic Command; 2nd Space Operations Squadron; and the services.

To determine the extent to which the Air Force had effectively developed and maintained the GPS IIIA integrated master schedule, we reviewed the program's schedule estimates and compared them with relevant best practices to determine the extent to which they reflects key estimating practices that are fundamental to having a reliable schedule. In doing so, we interviewed GPS program officials to discuss their use of best practices in creating the program's current schedule.

To assess the status of the GPS constellation, we interviewed officials from the Air Force Space and Missile Systems Center GPS program office, Air Force Space Command, and the 2nd Space Operations Squadron. To assess the risks that a delay in the acquisition and fielding of GPS III satellites could result in the GPS constellation falling below the 24 satellites required by the standard positioning service and precise positioning service performance standards, we obtained information from the Air Force predicting the reliability for 77 GPS satellites—each of the 31 current (on-orbit) and 46 future GPS satellites—as a function of time. Each satellite's total reliability curve defines the probability that the satellite will still be operational at a given time in the future. It is generated from the product of two reliability curves—a wear-out reliability curve defined by the cumulative normal distribution, and a random reliability curve defined by the cumulative Weibull distribution. For each of the 77 satellites, we obtained the two parameters defining the cumulative normal distribution, and the two parameters defining the cumulative Weibull distribution. For each of the 46 unlaunched satellites, we also obtained a parameter defining its probability of successful launch, and its current scheduled launch date. The 46 unlaunched satellites include 2 IIR-M satellites,[25] 12 IIF satellites, 8 IIIA satellites, 8 IIIB satellites, and 16 IIIC satellites; launch of the final IIIC satellite is scheduled for March 2023.

Using this information, we generated overall reliability curves for each of the 77 GPS satellites. We discussed with Air Force and Aerospace Corporation representatives, in general terms, how each satellite's normal and Weibull parameters were calculated. However, we did not analyze any of the data used to calculate these parameters.

Using the reliability curves for each of the 77 GPS satellites, we developed a Monte Carlo simulation[26] to predict the probability that at least a given number of satellites would be operational as a function of time, based on the GPS launch schedule approved in March 2009. We conducted several runs of our simulation—each run consisting of 10,000 trials—and generated "sawtoothed" curves depicting the probability that at least 21, 24, 27, and 30 satellites would still be operational as a function of time. We compared the results for a 24-satellite constellation with a similar Monte Carlo simulation that the Aerospace Corporation performed for the Air Force. We confirmed that our simulation produces results that are within about 2 percent of the Aerospace Corporation's results for all times between October 2008 and April 2024. Using 10,000 trials per run, the results of different runs of the same Monte Carlo simulation can vary by about 1 to 2 percent; consequently we concluded that we had successfully replicated the Aerospace Corporation's results. We then used our Monte Carlo simulation model to examine the impact of a 2-year delay in the launch of all GPS III satellites. We moved each GPS III launch date back by 2 years. We then reran the model and calculated new probabilities that at least 18, 21, and 24 satellites would still be operational as a function of time.

To assess impacts of a potential GPS service disruption on particular types of military and civil GPS users, we interviewed numerous military and civil GPS representatives and reviewed studies provided by civil agencies.

To assess the coordination and collaboration among federal agencies and the broader GPS community, and to determine the organization of the PNT community, we analyzed documents from and conducted interviews with officials in Washington, D.C. at the Office of the Assistant Secretary of Defense for Networks and Information Integration; SAF/USA (Air Force Directorate of Space Acquisitions); National Aeronautics and Space Administration; the Departments of Transportation, State, Commerce, and Homeland Security; the Space-Based National PNT Coordination Office; and the U.S. GPS Industry Council. We also interviewed a private sector GPS expert at Stanford University, and GPS industry representatives. To analyze how the U.S. government coordinates with foreign countries on GNSS (Global Navigation Satellite Systems), we met with representatives of and reviewed documents from the U.S. Department of State and European Space Agency (ESA) in Washington, D.C. To obtain information on efforts by Australia, China, Japan, and Russia to develop GNSS, we met with Department of State officials, reviewed materials provided by these countries' representatives at GNSS conferences, and consulted the official government space agency Web sites. We also traveled to Europe to meet with experts in satellite navigation at the European Space Agency, French Space Agency (CNES), European Commission Directorate-General for Energy and Transport Satellite Navigation Unit, and European GNSS industry experts. In addition, we attended a conference in Berlin, Germany to learn about international coordination on PNT systems and applications.

We conducted this performance audit from October 2007 to April 2009 in accordance with generally accepted government auditing standards. Those standards require that we plan and perform the audit to obtain sufficient, appropriate evidence to provide a reasonable basis for our findings and conclusions based on our audit objectives. We believe that the evidence obtained provides a reasonable basis for our findings and conclusions based on our audit objectives.

Appendix II: International Global Satellite Navigation Systems

In addition to the Global Positioning System (GPS), there are other space-based global navigation satellite systems (GNSS) in operation and in development. Russia has a system, GLONASS (Global Navigation Satellite System). There are currently 20 GLONASS satellites in orbit, and the Russians expect to have a full constellation of 24 satellites in orbit by 2010 and ultimately to expand to a 30-satellite constellation.

The European Union (EU) is developing its own GNSS program, Galileo. Originally started as a public-private partnership, the program now is completely funded by the public sector. The EU has 2 test satellites in orbit now, and plans to have a 27-satellite constellation with 3 spares by 2013. China also is in the process of developing its own GNSS, Compass (also called Beidou). China currently has 3 satellites in orbit, and plans to increase the constellation for coverage of the Asia-Pacific region by 2010 and for worldwide coverage by 2015. Table 5 lists the non-U.S. global navigation satellite systems currently in development.

Table 5. Non-U.S. Global Navigation Satellite Systems Currently in Development

System name	Country	Number of active satellites	Number of planned satellites	Planned date of full operation	Interoperable signals
Galileo	European Union	2 test satellites	27	2013	Interoperable L1C signal
GLONASS	Russia	20	30	2011	Interoperable L1C signal
Compass/ Beidou	China	3	35	Regional coverage in 2010	Compatible and interoperable with GPS, no broadcast on L1 C at this time

Source: GAO analysis based on information from foreign program presentations.

Appendix III: Cooperation between U.S. and Foreign Entities

During 2007, the Department of State signed joint statements of cooperation in the use of the Global Positioning System (GPS) with Australia and India. The Australia joint statement expresses the parties' intention to promote interoperability between GPS and Australia's Ground-based Regional Augmentation System and Ground Based Augmentation System. The India joint statement expressed the parties' intention to promote GPS and India's GPS and GEO-Augmented Navigation system. An executive agreement with the European Community and its member states has been in effect since 2004 that expresses the intention that GPS and Galileo will be interoperable at the user level for the benefit of civil users around the world. This cooperation has resulted in working groups that are reviewing technical, trade, and security issues. The technical issues described in the executive agreement involve GPS-Galileo radio frequency compatibility and interoperability and the design and development of the next generation of systems. For trade, a working group is determining how to maintain nondiscriminatory trade practices in the global market for goods and services related to space- based PNT, and a group was appointed to review the security issues concerning GPS and Galileo.

The United States and Russia initiated cooperation in 2004, with the parties expressing their intent to work together to maintain and promote civil interoperability at the user level between GPS and Russia's GLONASS system. Two working groups have been established to address: (1) radio frequency compatibility and interoperability for enhanced PNT and (2) technical interoperability between the search-and-rescue capabilities planned for GPS and GLONASS.

The United States and Japan have had a relationship since signing a joint statement in 1998. In the joint statement, the parties expressed their intent to promote and facilitate civilian uses of GPS. Japan is developing MTSATbased Satellite Augmentation System (MSAS), a geostationary satellite system similar to the U.S. Wide Area Augmentation System. The United States and Japan most recently met in November 2008 to discuss the civil use of GPS and Japan's MSAS and Quasi-Zenith Satellite System.

APPENDIX IV: COMMENTS FROM THE DEPARTMENT OF DEFENSE

OFFICE OF THE ASSISTANT SECRETARY OF DEFENSE
6000 DEFENSE PENTAGON
WASHINGTON, DC 20301-6000

APR 24 2009

NETWORKS AND
INFORMATION INTEGRATION

Ms. Christina Chaplain
Director, Acquisition and Sourcing Management
U. S. Government Accountability Office
441 G Street, NW
Washington, DC 20548

Dear Ms. Chaplain:

This is the Department of Defense (DoD) response to the GAO draft report, GAO-09-325, 'THE GLOBAL POSITIONING SYSTEM: Significant Challenges in Sustaining and Upgrading Widely Used Capabilities,' dated March 12, 2009 (GAO Code 120696).

The Department fundamentally concurs with the findings and recommendations expressed in the GAO report. However, factors relating to the longevity of the Global Positioning System (GPS) program and the complexity of its implementation across the range of military and civilian users require that our concurrence be augmented by clarifying comments, which have been provided separately.

GPS developmental satellites were initially launched over thirty years ago. The system has evolved successfully through at least three generations of spacecraft, ground control systems, and military user equipment. During that time, GPS has become one of the most widely used systems in the world for military and civilian positioning, navigation and timing (PNT) purposes and sets the example for other nations seeking to provide similar services.

In the United States, GPS is a fundamental enabler of national security and economic infrastructures, enhancing efficiency and improving safety and effectiveness of virtually all operations. As the cornerstone of our National PNT Architecture, it is the centerpiece around which future PNT services will evolve. The Department is fully aware of our responsibility with respect to GPS and is committed to maintaining and improving the services it provides. In that regard, DoD seeks the support of the Congress in maintaining stability of GPS funding to enable synchronized modernization of the next generation of GPS space, ground control, and user equipment that is now underway.

We have determined that the report is unclassified and has been cleared for open publication. Our point of contact for this review is Mr. Raymond Swider, (703) 607-1122, raymond.swider@osd.mil.

Ronald C. Jost
Deputy Assistant Secretary of Defense
(C3, Space & Spectrum)

Enclosure

GAO DRAFT REPORT DATED MARCH 12, 2009
GAO-09-325 (GAO CODE 120696)

"THE GLOBAL POSITIONING SYSTEM: SIGNIFICANT CHALLENGES IN SUSTAINING AND UPGRADING WIDELY USED CAPABILITIES"

DEPARTMENT OF DEFENSE COMMENTS TO THE GAO RECOMMENDATIONS

RECOMMENDATION 1: The GAO recommends that the Secretary of Defense appoint a single authority to oversee the development of the Global Positioning System (GPS) system, including space, ground, and user assets, to ensure that the program is well executed and resourced and that potential disruptions are minimized. (p. 43/GAO Draft Report)

DOD RESPONSE: Concur with comment. The Department has recognized the importance of centralizing authority to oversee the continuing synchronized evolution of the GPS. To that end, the Deputy Secretary of Defense has reaffirmed that the Assistant Secretary of Defense for Networks and Information Integration (ASD(NII)) is the Department's Principal Staff Assistant to oversee Positioning, Navigation, and Timing, and, specifically, is designated with authority and responsibility for all aspects of the Global Position System (GPS). This designation is contained in Department of Defense Directive (DoDD) 4650.05, issued on February 19, 2008. A formal Department of Defense Instruction is now in final coordination to further define the oversight processes to be employed in executing DoDD 4650.05, and completion is expected by May 2009. Further, under oversight of the ASD(NII), the U.S. Air Force is the single acquisition agent with responsibility for synchronized modernization of GPS space, ground control, and military user equipment. The Air Force acquires and operates the GPS space and control segments and provides the fundamental system design and security requirements necessary for acquisition of GPS user equipment and applications in support of diverse missions across the Department. Given the diversity of platforms, and equipment form factors involved, it is impossible for the Air Force to unilaterally produce a "one-size-fits-all" solution applicable to all DoD missions.

RECOMMENDATION 2: The GAO recommends that Secretary of Defense, as one of the Position Navigation and Timing executive committee co-chairs, address, if weaknesses are found, civil agency concerns for developing requirements and determine mechanisms for improving collaboration and decision making and strengthening civil agency participation. (p. 43/GAO Draft Report)

DOD RESPONSE: Concur with comment. The Department is aware that we employ a rigorous requirements process in support of our extensive operational and acquisition responsibilities and that the process is a source of frustration for civil agencies without similar processes in place. In an effort to address the issue, we have worked with the civil agencies to put in place a GPS Interagency Requirements Plan, jointly approved by the Vice Chairman of the Joint Chiefs of

Staff, who is in charge of our process, and the Department of Transportation (DOT), acting on behalf of all civil agencies. Further, we are now in the process of jointly coordinating the Charter for an Interagency Forum for Operational Requirements (IFOR) to provide meeting venues to identify, discuss, and validate civil or dual use GPS requirements for inclusion in the DoD GPS acquisition process. Finally, we sponsor educational outreach opportunities for civil agencies to become more fully acquainted with the DoD requirements process, including a day-long "Requirements Process Summit" jointly conducted by the Joint Staff and DOT on April 29, 2008. We will continue to seek ways to improve civil agency understanding of the DoD requirements process and work to strengthen civil agency participation.

End Notes

[1] For a list of reports on best practices, see Related GAO Products at the end of this report.

[2] GPS is augmented by ground-based or space-based navigation aids that are maintained by individual departments and agencies to provide users with improvements to the GPS navigation signal in terms of accuracy, availability, and/or integrity needs.

[3] GAO, *Space Acquisitions: Actions Needed to Expand and Sustain Use of Best Practices*, GAO-07-730T (Washington, D.C.: Apr. 19, 2007).

[4] GAO, *Best Practices: DOD Can Help Suppliers Contribute More to Weapon System Programs*, GAO/NSIAD-98-87 (Washington, D.C.: Mar. 17, 1998); *Space Acquisitions: Major Space Programs Still at Risk for Cost and Schedule Increases*, GAO-08-552T (Washington, D.C.: Mar. 4, 2008); and, *Defense Acquisitions: Results of Annual Assessment of DOD Weapon Programs*, GAO-08-674T (Washington, D.C.: Apr. 29, 2008).

[5] On July 31, 2004, the GPS program office became the GPS Wing, when the Air Force's Space and Missile Systems Center reorganized and renamed its organizations to mirror the traditional Air Force structure.

[6] Earned value management (EVM) is a program management tool that integrates the technical, cost, and schedule parameters of a contract. During the planning phase, an integrated baseline is developed by time-phasing budget resources for defined work. As work is performed and measured against the baseline, the corresponding budget value is "earned." Using this earned value metric, cost and schedule variances can be determined and analyzed. EVM provides significant benefits to both the government and the contractor. An EVM system is required on all DOD space-program-related contracts meeting certain thresholds unless waived by the DOD Space Milestone Decision Authority.

[7] The Defense Contract Management Agency (DCMA) is the DOD component that works directly with defense suppliers to help ensure that DOD, federal, and allied government supplies and services are delivered on time, at projected cost, and meet all performance requirements.

[8] GAO, *Space Acquisitions: Improvements Needed in Space Systems Acquisitions and Keys to Achieving Them*, GAO-06-626T (Washington, D.C.: Apr. 6, 2006).

[9] GAO, *Space Acquisitions: Major Space Programs Still at Risk for Cost and Schedule Increases*, GAO-08-552T (Washington, D.C.: Mar. 4, 2008).

[10] GAO-06-626T.

[11] Defense Science Board/Air Force Scientific Advisory Board Task Force, *Acquisition of National Security Space Programs*, Office of the Under Secretary of Defense for Acquisition, Technology, and Logistics (Washington, D.C.: May 2003).

[12] Defense Science Board Task Force, *The Future of the Global Positioning System,* Office of the Under Secretary of Defense for Acquisition, Technology, and Logistics (Washington, D.C.: Oct. 28, 2005).

[13] Independent Assessment Panel on the Organization and Management of National Security Space, *Leadership, Management, and Organization for National Security Space*, Institute for Defense Analysis (Alexandria, Va.: Jul. 15, 2008).

[14] Pub. L. No. 109-364 § 914.

[15] House Permanent Select Committee on Intelligence, *Report on Challenges and Recommendations for United States Overhead Architecture*, United States House of Representatives (Washington, D.C.: Oct. 3, 2008).

[16] The Program Objectives Memorandum (POM) is an annual memorandum submitted to the Secretary of Defense by the DOD component heads, which recommends the total resource requirements and programs within the parameters of the Secretary of Defense's fiscal guidance. The POM is a major document in the Planning, Programming, Budgeting and Execution process, and the basis for the component budget estimates. The POM is the principle programming document that details how a component proposes to respond to assignments in the Strategic Planning Guidance and Joint Programming Guidance and satisfy its assigned functions over the Future Years Defense Program. The POM shows programmed needs 6 years hence (i.e., in fiscal year 2004, POM 2006-2011 was submitted).

[17] This availability standard establishes thresholds for both global average and worst-case position dilution of precision (PDOP), a figure of merit commonly used to quantify the "goodness" of user-to-GPS-constellation geometry.

[18] Some users with more demanding requirements employ GPS augmentation systems that mitigate this impact. For example, many applications using augmentations such as Satellite-Based Augmentation Systems (SBAS), which in the United States is the Wide Area Augmentation

System (WAAS), have increased tolerance to degraded accuracy and availability when the constellation may be operating at minimum committed levels of availability.

[19] Both performance standards assume an unobstructed view of the entire sky except for 5 degrees above the local horizon.

[20] Defense Science Board Task Force, *The Future of the Global Positioning System*, Office of the Under Secretary of Defense for Acquisition, Technology, and Logistics (Washington, D.C.: Oct. 28, 2005).

[21] Antispoofing is a process of encrypting one of the codes broadcast by the satellites. This prevents an enemy from predicting the code sequence and using that prediction to generate a code that could be used to deceive a GPS set. The set would believe the deception code to be real and could falsely calculate its position.

[22] Geodetic refers to the use of geodesy for measurements. Geodesy is the science of measuring and monitoring the size and shape of the Earth.

[23] Augmentation systems are U.S. government global and regional systems that are maintained by individual departments and agencies to provide users with improvements to the GPS navigation signal in terms of accuracy, availability, and/or integrity needs.

[24] Although China is developing a space-based positioning, navigation, and timing system, the United States has not established a formal bilateral relationship with China. For the purposes of this report, "cooperative arrangements" is used to mean joint statements of cooperation and executive agreements.

[25] We completed our analysis prior to the successful launch of the first of these two IIR-M satellites on March 24, 2009.

[26] Monte Carlo simulation refers to a computer-based analysis that uses probability distributions for key variables, selects random values from each of the distributions simultaneously, and repeats the random selection over and over. Rather than presenting a single outcome—such as the mostly likely or average scenario—Monte Carlo simulations produce a distribution of outcomes that reflect the probability distributions of modeled uncertain variables.

In: Sustaining the Global Positioning System
Editors: Earl M. Peabody
ISBN: 978-1-60741-006-5

Chapter 2

GPS HEARING- HUBER TESTIMONY

Chet Huber

Thank you Mr. Chairman and good morning. I am Chet Huber, President of OnStar, a wholly owned subsidiary of General Motors Corporation. I am currently a member of the NASA PNT Advisory Board and have served on the CDC's Advisory Committee on Injury Prevention and Control.

With nearly 6 million active subscribers, OnStar is the leading telematics service provider employing over 2,200 individuals in the U.S. and Canada.[1] OnStar is now standard on virtually all General Motors' retail vehicles and has developed a prominent national brand position.[2]

Our core safety, security and peace of mind services include Automatic Crash Response and emergency services which we deliver from three call centers in Pontiac, Michigan; Charlotte, North Carolina and Oshawa, Canada. Other services include Turn-by-Turn navigation, stolen vehicle location; and monthly OnStar Vehicle Diagnostic emails. We also offer one button, hands-free, prepaid wireless calling.

In a typical month, after call screening, we provide unique and critical support for public safety agencies in responding to:

- Over 2,000 Automatic Crash Notifications and
- Over 10,000 occupant initiated Emergency Button presses. These include heart attacks, strokes, and crashes not triggering an automatic call. Last November, OnStar marked its 100,000th Automatic Crash Response.

Monthly, we also pass on to public safety:

- Over 6,000 Good Samaritan calls for everything from crashes involving other vehicles, to roadway hazards - to possible Amber Alert sightings. And we assist with
- Over 500 stolen vehicle location requests including - on many new models- the ability to slow a stolen vehicle down to avoid a high speed pursuit.

Other monthly service statistics include delivery of:

- over 3.4 million diagnostic emails,
- nearly one million Turn-by-Turn routes and
- Over 53,000 remote door unlocks.

Delivering these services, and growing to our current scale, has required deep and fundamental technological innovation as we've uniquely integrated cellular, GPS and voice recognition with extensive on-board and off-board software. This has required hundreds of millions of dollars of investments and resulted in the filing of over 500 patent applications with new filings at the rate of one every six days.

A critical element in our delivery of services is location. OnStar uses the civilian L1C/A signal to deliver our location-based services like Automatic Crash Response, stolen vehicle location assistance and Turn-by-Turn navigation. We also use directly or indirectly the GPS timing signal to enable other valuable services like remote door unlock and monthly diagnostic emails. An accurate, available and reliable GPS constellation is at the heart of our capability to provide these services.

We offer three recommendations for your consideration:

First, we must address the health of the current constellation. We are concerned that a recent report shows 8 of the current satellites are one component from total failure.[3] Loss of signal will likely immediately affect GPS accuracy and availability (geographic coverage).

Second, as the GPS system is modernized, it is imperative that the U.S. government formally commit to preserving the L1C/A signal and to ensuring backward compatibility for legacy applications with no loss of performance from current levels.

Automotive applications of GPS, like OnStar, are embedded into the vehicle's electrical system and subjected to extensive validation testing. Because of this, it is impractical to retrofit GPS related hardware and assure the reliable delivery of services to subscribers. Therefore, for the benefit of our millions of customers - and others facing similar legacy issues - we are asking Congress and the Executive branch to work together to develop a policy that supports backward compatibility at current performance levels.[4]

Regarding performance, it is important to understand that the current GPS system is performing at a level well above the specified minimum standard and operators have used that performance to improve and enhance services.

Any modernization initiative that degrades backward compatible performance - such as reducing the number of satellites making up the constellation - would likely adversely impact the provision of services by OnStar, including the quality of location information we provide to public safety - thereby potentially increasing the response time of public safety personnel to crash victims and others in need of emergency services.

Our third recommendation – and this is also important to legacy applications - is that we commit to maintaining the current PRN code (or satellite signature structure) for the primary orbital slots - as satellites in those slots are replaced. Legacy hardware is not capable of being expanded to accommodate more than 32 slots so renumbering above 32 will likely affect performance of legacy applications.

In conclusion, Mr. Chairman, OnStar and GM support the continued modernization of the GPS system. A modern and robust GPS system is essential to OnStar's ability to deliver future services.

Modernization will create opportunities to address current issues like urban canyons and to add new capabilities. For example, modernization offers the potential to support lane level accuracy in automotive applications, opening the possibility of important new vehicle-to-vehicle safety-related applications that are in experimental development. Additionally, strategies to support interchangeability with Galileo and other systems offer important opportunities to enhance performance.

But as we modernize, we must ensure backward compatibility at current performance levels for legacy applications like OnStar and thereby avoid significant disruptions in valuable services that carry important public safety benefits.

Thank you, Mr. Chairman. I look forward to your questions.

End Notes

[1] Telematics is a growing market. There is currently one additional significant telematics service provider with between 500,000 and one million active subscribers and several automotive manufacturers have announced plans to begin to offer telematics services in the future.

[2] OnStar also provides service to a small number of vehicles from other manufacturers.

[3] GPS Operations CGSIC Brief (Sep 15 2008), Civil GPS Service Interface Committee Meeting. The Constellation Status Summary additionally indicated that of the currently 29-32 healthy satellites, 19 are past design life and 18 are one component away from partial failure.

[4] In-vehicle screen based navigation systems are one example of applications that share similar legacy issues.

In: Sustaining the Global Positioning System
Editors: Earl M. Peabody
ISBN: 978-1-60741-006-5

Chapter 3

GPS Hearing- James Testimony

Larry James

Mister Chairman, Ranking Member Flake, and distinguished Members of the Security and Foreign Affairs Subcommittee, I am honored to be here today for my first opportunity to appear before you as United States Strategic Command's (USSTRATCOM) Commander of the Joint Functional Component Command for Space (CDR JFCC SPACE).

It's a distinct privilege to address you on the role we have in operating the Global Positioning System (GPS) constellation, and to represent the men and women of JFCC SPACE who employ space capabilities around the globe every day. These dedicated and innovative Soldiers, Sailors, Airmen, and Marines work hard to maximize the effectiveness of our space assets as they execute the command and control of our worldwide space forces.

Today I will focus my discussion on the importance of GPS to the warfighter, the health of our current GPS constellation, and USSTRATCOM's strategies to ensure the most robust space-based positioning, navigation, and timing (PNT) capabilities are provided by the GPS constellation.

Importance of GPS

GPS provides PNT data of such accurate and reliable nature to users worldwide that it has become essential to U.S. national security and economic well being. USSTRATCOM provides PNT effects to Department of Defense (DoD) global users that are critical to military operations. GPS is the centerpiece of global PNT services and the GPS constellation enables an ever-increasing arsenal of military and civil applications.

GPS provides critical services every second to our deployed forces around the globe from the infantrymen walking the streets of Fallujah, to the ships combating piracy off the coast of Somalia, to the aircraft patrolling our country's borders. The criticality of GPS to the warfighter is readily apparent in on-going operations in Afghanistan where GPS services provide foundational data, enabling us to track the location of U.S. and coalition forces.

For the military users, there are multiple examples of GPS successes. For example, precision accuracy of the GPS-guided Excalibur artillery rounds have enabled the U.S. Army to remove top al-Qaida operatives while firing within close proximity of infantrymen, reducing collateral damage and risk to Soldiers on the ground. In July 2007, with only two rounds fired, Excalibur took down a top al-Qaida target responsible for improvised explosive devices (IEDs), vehicle-borne IEDs, and indirect fire attacks. Excalibur's accuracy has also enabled the U.S. Army to engage targets in missions in the Diyala province of Iraq when aircraft were unable to provide support due to weather. On 27 March 2009, following an enemy ambush against a coalition unit, a B-1B Lancer used a GPS-guided 500 lb bomb to destroy a series of enemy fighting positions and a fortified heavy machine gun position near Tarin Kowt, Afghanistan. On 28 March 2009 near Asmar, Afghanistan, a formation of F-15E Strike Eagles attacked enemy forces with GPS guided Joint Direct Attack Munitions (JDAMs), hitting a sniper hideout and a group of enemy gunmen firing at coalition troops. Right now, the USS Eisenhower (Ike) Carrier Strike Group is on station supporting coalition forces executing OPERATION Enduring Freedom. On 26 April 2009, four F/A-18 Super Hornets flown from the deck of the Ike delivered four 500 lb GPS-guided JDAMs onto enemy fortified compounds and machine gun fighting positions, ending a fire fight with coalition forces. Clearly, the GPS constellation enables our forces worldwide to maneuver into a militarily advantageous position and then, through various GPS-aided munitions, exploit that tactical advantage to create effects ranging from tactical to strategic. GPS is critical to successful military operations across a multitude of engagements, and strong communication links, operational relationships, and reach back with the GPS Operations Center (GPSOC) and the Joint Space Operations Center (JSpOC) ensure USSTRATCOM is able to provide users the effects they need.

The User Operations Section within the GPSOC is the main interface for coordination of GPS effects with military users; however, the JSpOC also serves a vital role in military support. Whether it is a need for enhanced data accuracy, or the need to tailor additional operational support to a specific mission, the GPSOC and the JSpOC work directly with users to ensure warfighter needs are met. This means both the GPSOC and JSpOC are in constant communication with theater users to ensure optimum GPS availability and accuracy.

CONSTELLATION HEALTH AND GAP MANAGEMENT

As with all our military satellite constellations, the GPS constellation includes satellites which have exceeded their design life, operate with partial capability, or are a single key component away from failure. However, we have taken operational steps to mitigate the impacts of potential satellite losses to ensure continued support to warfighters and the global community of users.

Although required to maintain 24 GPS satellites on orbit at 95 percent probability, we have exceeded requirements by currently maintaining 30 operational satellites and have achieved sub-three meter accuracy. We conduct "residual operations" as an on-going solution to mitigate any potential gap in GPS by retaining older, partially mission capable satellites in a back-up mode that can potentially be returned to operations if a satellite in the constellation fails. Currently three vehicles are held in residual status, and through in depth analysis, residual satellites are returned to the constellation every six months to ensure PNT operational capability.

Additionally, we use "power management" as a way to extend the PNT mission on GPS satellites. While the planned degradation or suspension of secondary payloads are not regularly used as an operational tactic, technique, or procedure, we do regularly assess the health of individual satellites. As part of the assessment we balance the needs of the primary GPS mission, Space-based PNT, against the needs of the secondary missions on board. After coordination with the secondary payload mission team, those systems may be power managed or "shut down" to extend the PNT mission on the satellite.

While we continue to exceed current expectations and system requirements, we must continue to focus on future requirements for GPS capabilities for our military and civilian/commercial users. Ever-changing and increasing user requirements demand that our next generation of GPS incorporate a more robust anti-jam signal, more power directed capabilities, and faster signal acquisition to ensure users have the capabilities necessary to produce desired effects. Matching future user requirements with technological advances will allow USSTRATCOM to provide the most advanced and reliable space effects in response to the growing demands of the nation's warfighters and commercial users.

CONCLUSION

Space has a vital role, which is epitomized by the GPS constellation, across all instruments of national power – Diplomacy, Information, Military, and Economic interactions. Specifically, the United States' dependence on GPS across our military, civil, and commercial sectors requires PNT capabilities to ensure our ability to safely and effectively operate in diverse environments on the ground, at sea, in the air, and in space. Working in collaboration with other departments and agencies in the U.S. Government, the DoD must continue to build the relationships, processes, and capabilities within the global space community that allow us to operate effectively together to meet our national security objectives. You can be proud of your Soldiers, Sailors, Airmen, and Marines that expertly tackle the challenges we face every day.

In: Sustaining the Global Positioning System ISBN: 978-1-60741-006-5
Editors: Earl M. Peabody

Chapter 4

GPS HEARING- MCCASLAND TESTIMONY

Neil McCasland

Mr. Chairman, Ranking Member Flake, distinguished Members of the Subcommittee on National Security and Foreign Affairs, it is a privilege to address you on the Air Force's management and execution of the Global Positioning System (GPS) acquisition program. GPS provides accurate location and time information for unlimited users in all weather, day and night, anywhere in the world. It is vital to numerous military and civil activities, including mapping, aerial refueling, weapons delivery, rendezvous operations, search and rescue operations, banking, transportation and agriculture. The Air Force, as the developer, operator, and steward for GPS, is committed to maintaining GPS as the gold standard for positioning, navigation, and timing information. I will focus my discussion on recent program successes, plans to sustain and modernize civil and military capability, and the interagency partnerships required to guarantee GPS program success.

RECENT SUCCESSES

First, let me talk about a few recent successes. In September 2007, the 22-year-old, unsustainable legacy GPS Master Control Station was replaced by the Architecture Evolution Plan System, or AEP, which improved GPS accuracy, provided the capability to operate the GPS IIF satellites, and will increase the ability to protect the military's use of GPS signals. The transition from the legacy to this new control segment was seamless to GPS users worldwide.

We are also significantly modernizing the GPS space segment. In March we launched the seventh vehicle in the current generation of GPS satellites, known as Block IIR-M. This family of satellites supports a second civil signal, L2C, and a new fully-encrypted military-exclusive signal, M-code, that will provide improved anti-jam performance. The IIR-M launched in March just performed a successful test broadcast of a third civil signal, known as L5. This enabled the US to secure its filing for the frequency with the International Telecommunication Union.

The next generation of GPS satellites is the IIF family. This program has taken longer than planned to reach first launch due to the program's initial acquisition strategy, new requirements added for modernized GPS navigation signals and flexible power, defense industrial base consolidation, and technical issues. These issues propelled the Air Force to launch a campaign to rebuild our confidence in the schedule of this critical space system through a renewed focus on mission assurance.

We have now achieved multiple milestones for the first GPS IIF space vehicle including successful thermal vacuum and final integrated systems testing for satellite #1. At the same time, we are preparing to send the second space vehicle to Cape Canaveral as a pathfinder and also establishing a production line for the remaining vehicles. I'm confident we'll launch the first IIF in the first half of FY10, and the remaining space vehicles will be produced and available as needed over the next three to five years.

Finally, we have put great emphasis on the all-important user equipment. To date, the Air Force has procured nearly 345,000 Defense Advanced GPS Receivers, upgraded Miniaturized Airborne GPS Receivers, and procured 80,000 Ground Based GPS Receiver Applications Modules. Collectively, these deliveries represent significant enhancement to security and GPS anti-jam capability, increasing combat effectiveness.

PLANS TO SUSTAIN AND MODERNIZE

As important as these successes have been, we must continue to modernize. As we evolve all three GPS segments, we will deliver significant new capabilities. For the military user, both the GPS M-Code and increased power are key enablers of navigation warfare capabilities. The civil user will gain two new signals to provide greater accuracy, integrity, and utility.

A key element of our strategy is to deliver the first GPS III satellite to a known schedule. We worked hard with our requirements arm, our industrial partners, and our fiscal planners to ensure we integrated every lesson from the past to create a high-confidence GPS III schedule. In addition, our approach utilizes a significant amount of Lockheed Martin GPS IIR-M space vehicle heritage. The GPS IIIA program completed extensive pre-award risk reduction activities and demonstrated early designs and prototypes of higher-risk items. These efforts allowed us both to adopt a phased acquisition approach to pursue a low-risk set of capabilities for the initial satellites to support constellation replenishment and to allow longer technology maturation timelines for more challenging capabilities. Two weeks from now, the GPS IIIA program completes its Preliminary Design Review on schedule. This milestone demonstrates that the design addresses requirements with adequate margin to meet program objectives. The program is on track for a first launch in 2014.

The modernized control segment required to support the Block III Satellites, as well as critical information assurance and effect-based operations requirements, is currently in the technology demonstration phase and has completed system design review and prototype demonstration. We will soon down-select to one vendor to complete system development, with delivery in 2013.

New military user equipment is under development to exploit advances in the space and control segments. Currently, the three Military GPS User Equipment (MGUE) Technology Demonstration vendors continue to mature critical technologies, burn down risk, and provide a baseline for the new approach to information assurance requirements. All three vendors are developing prototype hardware capable of receiving new and existing GPS signals, and each has conducted a Critical Design Review. Prototype hardware deliveries begin in early Fiscal Year 2010. Based on the technology demonstration work, our approach to MGUE development and integration will emphasize modularity in the design to enable integration into the large number of end user applications and platforms. Available in 2015, these lighter, faster new receivers will be reprogrammable and include security and jam-resistant improvements.

The foundation for success both technically and programmatically in our modernization efforts lies in our mission assurance process. Mission assurance is a disciplined application of management, systems engineering, and quality principles over the entire lifecycle to ensure mission needs are satisfied. It is a culture that we've rebuilt at SMC that permeates the GPS team and is ingrained through all functional disciplines.

There are three elements are key to mission assurance culture: risk management, disciplined engineering, and a strong government oversight. Proper risk management has been in place from the beginning of the programs. Prior to awarding development contracts, the government conducted extensive risk reduction activity to establish an allocated requirements baseline and develop and demonstrate early designs and prototypes of higher risk items.

Disciplined engineering for GPS means we have taken a government-led enterprise approach to engineering processes. This approach emphasizes rigorous baseline management, institution of specifications and standards, and government responsibility for cost, schedule and technical trades. The enterprise approach also provides for integrated system-level testing to validate the service provided to warfighters and civil users alike.

Consistent with recommendations of several studies on acquisition, we've staffed the team with experienced and trained personnel who can exercise clear government oversight. Learning from the difficulties encountered with GPS IIF, we have placed responsibility to deliver our system back where it belongs – with the government. We have also put a team of retired military officers and senior contractor leadership in place to provide management, systems engineering, and business operations training and mentoring for these personnel.

Simultaneously, senior leadership across the Air Force, DoD, and DoT have committed to GPS program success. This shared goal enhances capability synchronization, budget advocacy and stability, and provides us the support we need to deliver to our plan.

UNITY OF EFFORT

GPS supports both military and civil users and requires involvement, advocacy, and resources from the DoD and DOT. Within the DoD, Air Force Space Command is responsible for leading GPS requirements development and advocacy through the Joint Requirements Oversight Council. Acquisition responsibility for GPS falls under the Program Executive Officer for Space working through the Deputy Undersecretary of the Air Force for Space Acquisition and the Undersecretary of Defense for Acquisition, Technology and Logistics.

DOT is the lead civil agency for GPS and coordinates the requirements, resources, and advocacy across the civil community. To ensure unity of effort between DoD and DoT, the President signed the US Space-Based Positioning, Navigation, and Timing Policy on 8 December 2004. This policy has established guidance and the implementation actions for the space-based PNT programs, augmentations, and activities for US national and homeland security, civil, scientific, and commercial purposes. The policy also created a National Space-Based PNT Executive Committee to advise and coordinate among federal agencies. This process is working well, and has yielded great benefit for the GPS enterprise.

CONCLUSION

The Air Force is committed to maintaining GPS as the premier provider of positioning, navigation, and timing for users around the world. We have high-confidence plans to sustain and modernize this critical national capability on planned schedules.

In: Sustaining the Global Positioning System
Editors: Earl M. Peabody
ISBN: 978-1-60741-006-5

Chapter 5

GPS Hearing- Swiek Testimony

F. Michael Swiek

Thank you Chairman Tierney, Ranking Member Flake, and distinguished members of this Subcommittee, for providing an opportunity today to discuss this important topic.

The Global Positioning System (GPS) is one of the great U.S. success stories involving a shared national asset. GPS is a national model of successfully balancing military advantage and civilian equities to serve a broad and diverse range of national interests, from national security and public safety, to enabling critical infrastructures, advancing scientific research, facilitating local government productivity, and enhancing the productivity and competitiveness of diverse industries that are important to our economy and serving millions of individual Americans every day.

GPS is a model of government-industry cooperation contributing to the national economy through the entrepreneurial creation of companies, industries and jobs deriving value for users from integrating GPS positioning, navigation, and timing information in application solutions. The initial military investment in GPS has not only met military requirements and demonstrated invaluable military utility for the warfighter; it also provided a signal for civilian use at little to no additional cost. The dedicated men and women at USAF Space Command have achieved superb operational management of the GPS constellation for all users under Air Force stewardship of GPS. This operational excellence, together with predictable U.S. policy over two decades, has given the global community a stable, reliable signal that has provided a solid foundation for tremendous private sector investments in receiver and applications innovation. The result of this enlightened U.S. approach has been the worldwide adoption of GPS as a global information utility, providing major productivity benefits to the Nation.

It is difficult to say with precision just how big the GPS industry is because it touches and contributes to so many different applications and areas in so many ways. Some recent estimates have impressive numbers such as $ 15-50 billion annually worldwide depending on how one counts direct and indirect effects. GPS is a core information technology for many industries that are key to the U.S. economy. Examples include agriculture, aviation, construction, vehicle navigation, fleet management, public safety, geographical information systems, land use, environmental monitoring, earthquake monitoring, wildlife monitoring, disaster management, telecommunications, E911 cell phones, mapping, mining, marine transportation, surveying, infotainment, and probably a couple of hundred additional areas or more.

More impressive than the aggregate value of U.S. or worldwide GPS industries is the effect that GPS can have on the productivity and competitiveness of key industries. GPS enhances productivity at times as much as 30 percent through exploitation of precise positioning, navigation, and timing in value chains. It is not an exaggeration to say that GPS is everywhere, not only where we commonly and almost ubiquitously see it, such as in consumer car navigation devices. It is there, essential and critical, even where you don't realize it. Whenever you make a call on your cell phone, withdraw money from your ATM, or send an e-mail, you are using GPS. GPS precise time signals are essential tools for synchronizing the networks through which these services operate. Turn on a light, and you are probably using GPS as well, as electric power grids similarly use GPS precise time signals for synchronization and fault detection. The roads you drive on may have been built by construction equipment guided by GPS. Not only has the term "GPS" become a common term in the public lexicon, it has become an essential and critical utility on which public and private infrastructures depend.

U.S. industry has been a major factor and leader in the development of today's GPS industry through entrepreneurial vision, technological innovation, and private sector investment. But, we have not done this alone. The U.S. Government has promoted and encouraged this development by establishing, maintaining and reinforcing a stable policy framework that has consistently received farsighted and bipartisan support. It has been a true partnership of shared visions, discussions and debates, cooperation and coordination. This has been possible through the open dialogue that has taken place since the early days of

GPS, some 25 plus years ago, between civilian and military, industry and government on technical and policy issues as the technology, system and applications have evolved.

As we move forward to new generations of GPS satellites and signals, the challenge is to maintain this impressive level of reliability and stability. Successful adoption of modernized civilian GPS signals will occur if the installed user base can continue to trust the consistent and stable policy framework that the U.S. Government has provided for GPS for two decades. The new signals will need to sustain the legacy of accuracy, availability, and reliability established over the past twenty years. The adoption of GPS is a testament to the trust of users in Air Force stewardship. Users rely on the ability of the Air Force to operate and maintain the satellite constellation and stable signal structures that serves the warfighter and diverse civilian users in a way that enhances both our national and economic security. We strongly encourage the continuation of the open and balanced dialogue between all stakeholders; users and providers, civilian and military, industry and government. Our industry association strives to be an objective information resource to support that dialog.

Thank you. I would be happy to answer any questions you might have.

Commercial GPS Application Examples

The following examples of GPS applications provide user testimony that describes their decision metrics (e.g., productivity gains, return on investment and timeframe) to invest in and use commercial and consumer GPS products:

Precision Survey

- **California:** The Bay Bridge replacement project benefited from precision GPS and optical technologies. The new 6,500-ton, 350-foot section of the San Francisco-Oakland Bay Bridge, used precision GPS to "retrofit-by-replacement" an opening in the existing bridge. One of the busiest bridges in the U.S. was shut down on August 31, 2007 at 8:00 pm and seventy hours later—11 hours ahead of schedule—the bridge was reopened.

- **Chicago, Illinois**: Rebuilding the Dan Ryan:

One of the busiest expressways in the U.S., Chicago's 14.5-km Dan Ryan Expressway underwent complete reconstruction from 2005 to 2007. The largest expressway rebuild in Chicago history, the massive $975-million project stayed on its swift three-year timeline by utilizing the latest GPS-based surveying, 3D data modeling, and machine control technology.

Divided into five sections, the project was a model of efficiency, quality control and productivity. By the end of 2007, the rebuilt road included several new lanes, reconfigured ramps, a new interchange, and enhanced lighting and sewers. The reconstructed Dan Ryan is safer, wider, more durable and easier to access.

The project's sections were connected through a precise control network. Set by the engineering companies using GPS-based systems, the control network covered 12,950 km2 and facilitated project coordination. In addition, the local precision GPS network was used by project surveyors who wanted to connect to the network without any base station.

For earthwork and construction, the general contractor and subcontractor used GPS grade control equipment, real-time precision GPS surveying systems and optical total stations. With grading, slope and actual position information inside the cab, they could run machine crews around the clock, greatly increasing productivity, minimizing errors and streamlining costs. They estimated that by project end about a million cubic yards of dirt was moved.

The earthwork crews were connected to the larger project through a 3D road model calibrated to the precision GPS network. The road model was downloaded in real time into the GPS-based control equipment and calibrated to within 9 mm tolerance.

Using that same 3D road model, the survey subcontractor's four crews utilized GPS rovers and total stations to set over 91,440 linear m of construction survey layout, as well as perform other tasks. The road model allowed the crews to work quickly and accurately, keeping up with around-the-clock schedule.

The Dan Ryan's fast pace and high accuracy requirements highlight how GPS-based technology helps connect the jobsite for greater efficiency, accuracy, and productivity.

Precision Agriculture

Small growers to agri-business throughout the U.S. and worldwide use high precision GPS products in every step throughout the growing process.

- **Nebraska:** precision GPS network expands to meet increasing demand for +/- one inch repeatable accuracy:

  In 2004, a company operating this precision network added capacity to meet fast growing customer demand for +/- one inch repeatable accuracy of precision GPS autopilot systems for farm equipment use in precision applications

  In 2006, this company expanded capacity to serve 35 customers throughout their trade area

  Some of the precision network signals are reliable as far as 12 miles from the transmitter (serving about 129,000 to 163,000 acres).

  The customer may move the precision GPS receiver to different machines throughout the year" according to the company president.

- **Illinois** corn growers see big payback on precision GPS steering systems:

  This corn grower raises about 3,000 tillable acres of corn and soy beans. "These precision GPS systems save us $17,000 per year in fuel, labor, and repairs. At the same time, they increase our income by $54,000 each year due to less compaction, precise placement of seed over strip-tilled fertilizer, plus increased productivity and timeliness of planting and harvest operations. And those savings are based on $2.50 per bushel of corn and $2.50 per gallon of diesel fuel."

  "These reduced costs and extra income from precision GPS products are repeatable—they happen every year and provide a return on investment of 110 to 120 per cent per year." The longest payback for a customer is approximately 2.5 years and the fastest payback is 3 months.

- **Kansas** grower:

  "KSU research shows a possible 10-15 percent yield increase with strip tillage. Last Fall, we decided to upgrade to automated steering with precision GPS. When we knifed in Fall anhydrous, our pass-to-pass accuracy was one inch or better—often ¼ to ½ inch. When I went back with the N applicator, our system dropped the knives right back into the same grooves. This spring we will plant with our GPS on a Cat 55, then put the system on our sprayer. In a year or two, I can see adding a second GPS system, so we can run it on different jobs at the same time."

Precision Construction

- Big Island, **Hawaii:** development project is being built on a 450-acre site, mindful of the delicate ecosystem, shoreline offset requirements, and indigenous culture:

  The construction company added GPS-based automatic grade control in 2005.

  The fact that operators have instantaneous cut-and-fill feedback is something they like. When they are ripping, they can drop their blades and the display will say you need to drop the blade another two feet to be on plan.

  A significant advantage of the GPS-based automatic grade control system is that the work can be accomplished in one pass with little need for rework.

  "When this huge, complicated project is complete, I will be able to say that the finished grades match exactly what is on the plan with no deviations or exceptions—including the golf course, which understandably has to be dead on."

  "The feedback from the developer is that we're hitting grade consistently within three-tenths of an inch of the finished contour. They are amazed that we're grading that close to the plan specs."

  "Without automatic grade control, it would have been a nightmare to maintain accuracy on the slope. Because the dozers have the site plan right in the cab, we save a tremendous amount of time and increase our productivity."

- **Georgia:** Construction project: 2 million square foot footprint of a logistics warehouse:

  "We're glad our motor grader is equipped with the latest GPS-based grade control, according to the construction company building this project. We wouldn't have been nearly as precise, nor completed the pad as quickly.

  "We had our own survey crew check the building pad when it was completed. The finished pad was consistently within a half inch of the plan, throughout the whole plan throughout the whole expanse, fully one-third of the mandated tolerances... which was very, very, impressive."

  According to the construction company, the right tolerances of the graded pad made for much smoother placement of concrete. If you are even a quarter-inch on two million square feet, that is a lot of additional concrete. It is estimated that each additional quarter-inch of concrete would cost $1 million dollars."

"When you're talking about the major advantages of equipping our motor graders and dozers with the GPS-based systems, the big on is time....it takes less time to do the work and to do it right. We're completing projects sooner, and we've literally cut our rework by 70 percent using the GPS system."

As an example of the time saved, the construction company pointed to a four acre section of the parking lot that they had recently graded in 2 and ½ days. "If I had completed that section the conventional way with driving hubs every 25 feet, it would have taken six days."

- **Wisconsin:** a smaller excavating and utilities contractor:

  "You know we're still relatively new to the GPS-based grade control technology and the site positioning system rover. What sold my management were the productivity gains, combined with the desire to remain competitive in this market where bids can be separated by pennies per yard."

Fleet Management in Service to the Elderly and Handicapped

Santa Clara County, **California**. As a non-profit, public benefit organization, OUTREACH is committed to the mission of supporting older adults, individuals with disabilities and low-income families with their efforts to lead independent and self-sufficient lives. From our origins as a War on Poverty program over 30 years ago, OUTREACH has grown to be one of the nation's premier Adult with Disability Acts (ADA) Paratransit and community transportation providers. OUTREACH's service model is unique because it combines human service values with a transportation system that incorporates cutting-edge technologies, such as the Global Positioning System (GPS) and custom software solutions to increase program efficiency and cost effectiveness. The first year of operating the GPS-based paratransit vehicle tracking system, Outreach saved approximately $435,000.

In: Sustaining the Global Positioning System ISBN: 978-1-60741-006-5
Editors: Earl M. Peabody

Chapter 6

GPS HEARING- VAN DYKE TESTIMONY

Karen Van Dyke

Chairman Tierney, Ranking Member Flake, and Members of the Subcommittee:

I am Karen Van Dyke, Acting Director for Positioning, Navigation and Timing in the U.S. Department of Transportation's Research and Innovative Technology Administration (RITA). I appreciate the opportunity to appear before you today to discuss the criticality of the Global Positioning System to the civil user community.

GPS technology is increasingly woven into the fabric of American society, from cars and planes to cell phones and wristwatches. It improves productivity and efficiency in many areas of commerce. For example, today's construction, farming, mining, shipping, surveying, and traffic management systems have become dependent on GPS. The technology enhances public safety by preventing transportation accidents and by reducing the response times of ambulances, firefighters, and other emergency services. It allows agriculture operations to continue through low visibility field conditions such as rain, dust, fog and darkness, and to apply chemicals precisely, reducing environmental impact while reducing production costs. GPS also furthers scientific aims such as weather forecasting, earthquake prediction, and environmental protection.

Furthermore, the precise GPS time signal, derived from atomic clocks, is embedded in critical economic activities such as synchronizing communication networks, managing power grids, and authenticating electronic transactions.

Importance of GPS to NextGen

Of particular interest to the Department of Transportation is the Federal Aviation Administration's (FAA) Next Generation Air Transportation System (NextGen) program. NextGen is a wide-ranging transformation of the national air transportation system to meet future demand and support the economic viability of the system while reducing delays, improving safety, and protecting the environment. NextGen will change the way the system operates – reducing congestion, noise, and emissions, expanding capacity and improving the passenger experience. NextGen is a highly complex, multilayered, evolutionary process of developing and implementing new technologies and procedures.

NextGen will reduce fuel burn and greenhouse gas emissions, allow more direct, time-based routings, enable safer operations, and reduce runway incursions. United Airlines already has pioneered the use of tailored arrivals based on GPS from Honolulu to San Francisco, with a fuel savings of 1,600 pounds per flight.

GPS is the foundation for NextGen navigation and surveillance. The continuity of funding and integrity of the planned launch schedule of the GPS constellation is vital to the nation moving ahead with NextGen.

Commitment to GPS

I would like to thank the Air Force for their dedicated service in providing extremely reliable operation of GPS since it achieved Initial Operating Capability in 1993. The United States clearly is the leader in space-based positioning, navigation, and timing and we must continue to maintain and improve GPS, its augmentations, and backup capabilities to meet growing national security, homeland security, economic security, civil, and scientific demands, and to maintain this U.S. technology leadership position.

Sustainability of the GPS constellation is critical to users worldwide. The Department of Transportation is committed to modernization of GPS and providing funding to ensure the development and modernization of the next generation of GPS to provide new civil capabilities. Fully funding the DOT portion of GPS modernization is critical to ensuring that the GPS III program remains on schedule to ensure future constellation sustainment.

The Department of Transportation is confident that the Department of Defense will continue to operate GPS at or above the minimum GPS Performance

Standard commitment of 21 healthy satellites 98 percent of the time, equivalent to 24 healthy satellites 95 percent of the time and will find innovative methods to extend the life of the GPS satellites to prevent any gaps in availability. We recognize that GPS has exceeded performance commitments with 30 satellites currently operational, and that some users may have come to expect this level of service.

Mitigation of Disruption

The Department of Transportation is a provider, as well as a user, of GPS services, augmenting the GPS signal to improve accuracy and integrity. FAA provides the Wide Area Augmentation System (WAAS), and RITA coordinates resources and plans for the inland component of the Nationwide Differential GPS System (NDGPS), operated and maintained by the U.S. Coast Guard. WAAS and NDGPS stations are a part of the National Oceanic and Atmospheric Administration (NOAA)-managed national Continuously Operating Reference Stations or CORS network of over 1300 permanently operating GPS receivers maintained by over 200 federal, academic and private organizations. The U.S. Air Force, U.S. Coast Guard, and the Federal Aviation Administration have agreements to coordinate and provide notification of GPS performance and any disruptions of GPS service to the user community.

For aviation users relying on unaugmented GPS, when the constellation is at its minimum GPS Performance Standard commitment, outages will be experienced on a routine basis, which could result in complaints and economic impact. For users who equip with GPS augmented by WAAS, the impacts are reduced, supporting minimum availability requirements of 99% or more.

However, like any radionavigation system, GPS is vulnerable to interference that can be reduced, but not eliminated. In 2001, RITA's Volpe National Transportation Systems Center issued the "Vulnerability Assessment of the Transportation Infrastructure Relying on the Global Positioning System". The findings of this assessment indicated that there was awareness within the transportation community of risks associated with use of GPS as a primary means for position determination and precision timing. Due to the reliance of transportation on GPS signals, it is essential that threats be mitigated and alternative back-ups be available, and the system be hardened for critical applications. DOT has determined that sufficient alternative navigation aids currently exist in the event of a loss of GPS based services.

Potential back-up capabilities to GPS are being explored as part of a National Positioning, Navigation and Timing (PNT) Architecture study, initiated in 2006 at the request of the Assistant Secretary of Defense for Networks and Information Integration and DOT's Under Secretary of Transportation for Policy. The overarching goal of the architecture, with GPS as its cornerstone, is intended to overcome identified capability gaps, and achieve an evolutionary path to providing integrated space-based, terrestrial, and autonomous solutions in the 2025 time period that will ensure the continuity of government-provided PNT services.

In conclusion, I would like to thank the Committee for allowing me to discuss the civil user perspective of GPS. The Department of Transportation is committed to continue our strong working relationship with the Department of Defense to maintain our global leadership in space-based PNT.

I would be glad to answer any questions you may have.

Chapter Sources

The following chapters have been previously published:

Chapter 1 – This is an edited, excerpted and augmented edition of a United States Government Accountability Office (GAO), Report to the Subcommittee on National Security and Foreign Affairs, Committee on Oversight and Government Reform, House of Representatives. Publication GAO-09-325, dated April 2009.

Chapter 2 – These remarks were delivered as Statement of Mr. Chet Huber, President, OnStar, before the Subcommittee on National Security and Foreign Affairs, Committee on Oversight and Government Reform, House of Representatives, dated May 2009.

Chapter 3 – These remarks were delivered as Statement of Lieutenant General Larry James, Commander, Joint Functional Component Command for Space, before the Subcommittee on National Security and Foreign Affairs, Committee on Oversight and Government Reform, House of Representatives, dated May 7, 2009.

Chapter 4 - These remarks were delivered as Statement of Major General Neil McCasland, Director, Space Acquisition, Office of the Secretary of the Air Force, before the Subcommittee on National Security and Foreign Affairs, Committee on Oversight and Government Reform, House of Representatives, dated May 7, 2009.

Chapter 5 – These remarks were delivered as Statement of Mr. F. Michael Swiek, Executive Director, U.S. GPS Industry Council, before the Subcommittee on National Security and Foreign Affairs, Committee on Oversight and Government Reform, House of Representatives, dated May 7, 2009.

Chapter 6 - These remarks were delivered as Statement of Karen Van Dyke, Acting Director, Positioning, Navigation, and Timing Research, Research and Innovative Technology Administration, U.S. Department of Transportation, before the Subcommittee on National Security and Foreign Affairs, Committee on Oversight and Government Reform, House of Representatives, dated May 7, 2009.

INDEX

N

O

P

Q

R

S

T

U

V

W

Y